Handbook of OMT Review

Sixth Edition

Handbook
of

OMT REVIEW

by Lori A. Dolinski, MSc, PhD, DO

Copyright 2010

Pro-Medica Publishing Company

ISBN 978-0-557-76433-4

Handbook of OMT Review
Sixth Edition

Library of Congress Cataloging-in-Publication Data
Handbook of OMT Review/Lori A. Dolinski, MSc, PhD, DO-6th ed.
ISBN 978-0-557-76433-4
1. Handbook of OMT Review   II. Dolinski, Lori A.

HANDBOOK OF OMT REVIEW

Written by Lori A. Dolinski, MSc, PhD, DO
Illustrations by Lori A. Dolinski, MSc, PhD, DO
Printed in the United States of America
Printed and Distributed by Lulu, Inc.
Published by Pro-Medica Publishing Company, Revere, PA 18953

HANDBOOK OF OMT REVIEW

TABLE OF CONTENTS

Chapter One: **INTRODUCTION**

<u>Somatic Dysfunction</u> - the impaired or altered function of the somatic system, including the skeleton, joints, and myofascial structures as well as associated neural, vascular, and lymphatic elements.

<u>Diagnostic Criteria</u>

Tissue texture changes (such as edema, erythema, coolness, ropiness, moisture)

Asymmetry (as in lack of symmetry from one side to the other of bones, joints, muscles, and related structures)

Restriction (as in restricted movement)

Tenderness

Acute versus Chronic Somatic Dysfunction

Diagnostic Feature	CHRONIC	ACUTE
Tissue Texture Change	Pale, fibrotic, ropy, stringy, cool, flaccid, dry, scaly, pimply, contractured, with hypotonic muscles	Erythematous, boggy, spongy, edematous, hot, moist, with hypertonic muscles
Asymmetry	+	+
Restriction	+ (with little to no pain on movement)	+ (with pain on movement)
Tenderness	Achy, dull, burning, with paresthesias, gnawing	Sharp, intense, severe, stabbing, throbbing

<u>Types of Movement</u>
ACTIVE - patient moves himself in a given way.
PASSIVE - patient relaxes while another person moves that patient in a given way.

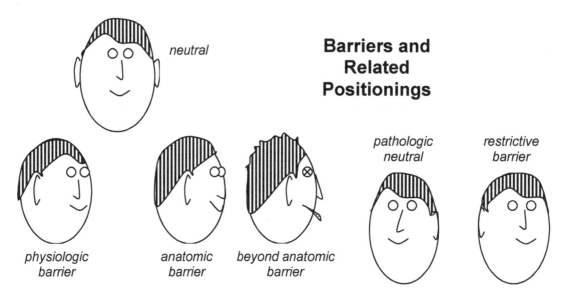

neutral

**Barriers and
Related
Positionings**

*physiologic
barrier*

*anatomic
barrier*

*beyond anatomic
barrier*

*pathologic
neutral*

*restrictive
barrier*

Restriction of Movement and Barriers

Physiologic barrier - point to which a patient can *actively* move a particular joint

Anatomic barrier - point to which a patient can *passively* move a particular joint

Restrictive (or Pathologic) Barrier - the barrier in pathologic situations that exists before the physiologic barrier. In other words, the patient is not able to move his joint as far as is normally possible and the motion is said to be restricted. The actual area of restriction exists after the point of the pathologic/restrictive barrier.

Fryette's Principles
  ➤ applies to thoracic and lumbar spine only
  ➤ DOES NOT apply to cervical or sacral spine
  ➤ "**one away, two together**" ➔ Fryette 1 sidebending and rotation occur "away" from each other (i.e., in opposite directions), while Fryette 2 sidebending and rotation occur "together" (i.e., in the same direction).

TYPE 1 FRYETTE (Principle 1)

o Neutral
o Sidebending and rotation occur in opposite directions
o In nomenclature, sidebending precedes rotation, i.e. NS_RR_L In motion mechanics, sidebending precedes rotation
o Usually involves a group of vertebrae, i.e. T8-10 NS_RR_L

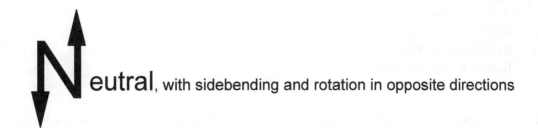

Neutral, with sidebending and rotation in opposite directions

TYPE 2 FRYETTE (Principle 2)

o Flexed or extended (non-neutral)
o Rotation and sidebending occur in the same direction
o In nomenclature, rotation precedes sidebending, i.e. FR_LS_L. In motion mechanics, rotation precedes sidebending
o usually involves one vertebra, i.e. T7 FR_LS_L

Extended/ **F**lexed

TYPE 3 FRYETTE (Principle 3)

o Motion of any vertebral segment in one plane has an effect on the amount of mobility of that segment in other planes.

Nomenclature
1. The dysfunction is named for the freedom of movement.
2. Restriction of motion of a segment is described in terms of the restricted segment(s) as it is above the functional segment, i.e. T10 is extended, rotated and sidebent to the right *on* T11.
3. And, of course, the different Fryette types list the sidebending and rotational components in different order.
4. There is no need to record the subscripts of the 2nd planar component when describing any dysfunction obeying Fryette principles, i.e. $L1NSR_R$ is acceptable notation since it is automatically assumed to be $L1NS_LR_R$ (because of our understanding of Fryette principles)

Quickly Evaluating Somatic Dysfunction in the Thoracic and Lumbar Spine
    Check for these 3 components in this order:
    1.   Rotation in neutral
    2.   Rotation in flexion
    3.   Rotation in extension

Example:
    1. Checking rotation in neutral: (illustration is a view as if one were looking down with X-ray vision)

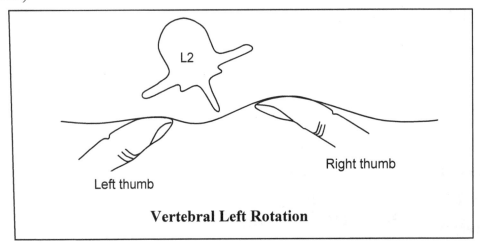

Vertebral Left Rotation

a. The right finger can go in deeper than the left and the left finger is locked by the transverse process; this is because the vertebra is rotated left. The left thumb is described as more posterior than the right

b. So, the vertebra is rotated left, and is unable to properly rotate right.

Therefore, the restriction is in rotating right and the freedom of movement is found in rotation to the left (the way the vertebra is freely going). Therefore, this vertebra is, so far, known to be an L2 rotated left.

    2. Check rotation in flexion.
    3. Check rotation in extension.

In the end, record the dysfunction in terms of the freedom of movement for all planes: neutral/flexion/extension with rotation and sidebending. For example, L2-3NS$_R$R$_L$ or L2ER$_R$S$_R$ or L2FR$_R$S$_R$.

Which is it .....**neutral, flexed, or extended??** Again, name the dysfunction for the ease of movement, which is best represented in this case by that position which most permits "normal" segment alignment. For example, let's say T1 is rotated right. If that rotation increases with flexion, remains rotated right in neutral, but returns to a more neutral position with extension, then T1 will be noted as being extended, or "E".

Planes of Motion

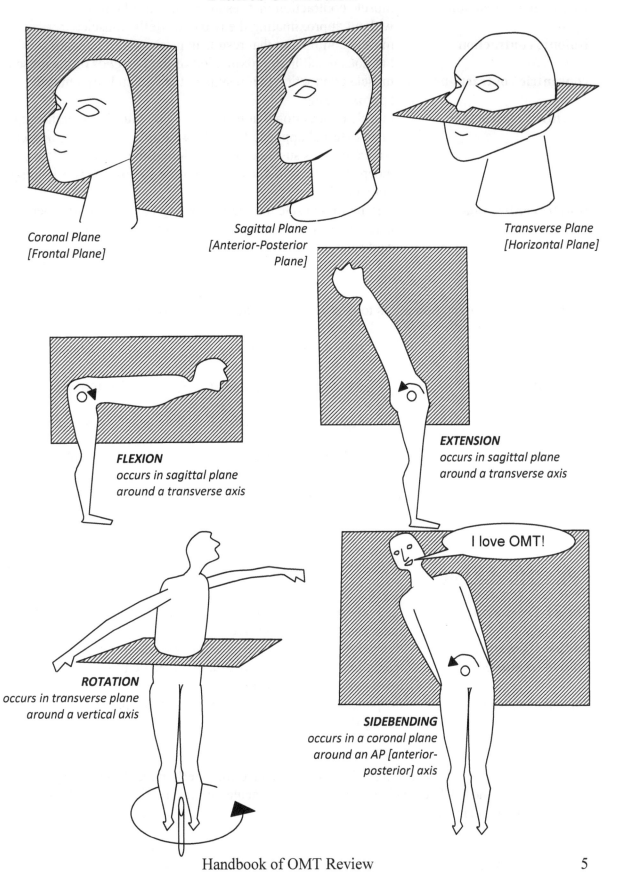

Coronal Plane
[Frontal Plane]

Sagittal Plane
[Anterior-Posterior
Plane]

Transverse Plane
[Horizontal Plane]

FLEXION
*occurs in sagittal plane
around a transverse axis*

EXTENSION
*occurs in sagittal plane
around a transverse axis*

ROTATION
*occurs in transverse plane
around a vertical axis*

I love OMT!

SIDEBENDING
*occurs in a coronal plane
around an AP [anterior-
posterior] axis*

Muscle Contraction

isometric contraction
muscle contraction that results in increased tension without approximating the two ends of the muscle.

isotonic contraction
muscle contraction that results in no increased tension, but does result in approximation of the two ends of the muscle.

concentric contraction
muscle contraction that results in the two ends of a muscle approximating.

eccentric contraction
muscle contraction that results *with* that muscle's lengthening due to an external opposite force. It is a form of isotonic contraction. An example is sustained contraction of the biceps while the biceps lengthens to allow one to place a book one was holding onto a table.

isolytic contraction
forcing the muscle to lengthen during contraction. It is neither isometric or isotonic. An example is the loser in an arm wrestling contest.

Facets

The following orientation refers to the SUPERIOR facets: "Bum-Bul-Bm"

Acronym	Direction the Facets face	Region
BUM	backward, upward, medial	Cervical
BUL	backward, upward, lateral	Thoracic
BM	backward, medial	Lumbar
	--no facets--	Sacral

TREATMENT BASICS

Treatment Approaches:

Direct Treatment:
the body/body part is moved toward the restrictive barrier; the barrier is said to be engaged

Indirect Treatment:
the body/body part is moved away from the restrictive barrier

Active treatment:
the patient assists in the treatment; the patient is "actively" involved with the movement

Passive Treatment:
the patient relaxes during the treatment; the patient is "passive" during the movement

Clinical Note:

1. Direct treatment is generally NOT recommended for elderly patients, chronically ill patients, hospitalized patients, or patients with acute injuries, particularly acute neck injuries of any kind.

2.   HVLA is NOT recommended for patients with the potential for easy fractures, such as patients with osteoporosis, osteogenesis imperfecta, bone cancer, and other cancers with metastatic potential to bone. In fact, all of the aforementioned disorders represent absolute contraindications to the use of HVLA. More absolute contraindications include severe rheumatoid arthritis (with respect to cervical HVLA), patients at risk for vertebral artery dissection (with respect to cervical HVLA), any kind of cancer in the vicinity of the intended treatment site, the use of cervical HVLA on Down's syndrome patients, and fractures in the area of treatment. Relative contraindications are many, and include acute whiplash, pregnancy, post-op state, herniated disc, hemophilia, acute sprain at the intended treatment site, current anticoagulant use, and any type of vertebral artery disease (with respect to cervical HVLA).

3.   Lymphatic techniques are NOT recommended for more advanced cancer patients.

Interval Between Treatments
INCREASED in sicker patients, elderly, most chronic cases
DECREASED in pediatrics patients, acute cases

Sequence of Treatment
No matter where on this schematic the pain is perceived, all treatment should start at the beginning of the schematic, i.e. at the most central rather than the more peripheral location. Once treatment is begun in that central location, the practitioner then continues to each subsequent region as dictated by the flowchart until the area with complaint is reached, i.e. generally the practitioner works from central to peripheral locations. The exception is in the case of very acute somatic dysfunction; in these situations, peripheral areas are treated first to allow one eventual access to the central area.

| Central | | Peripheral |

Upper thoracic spine → upper ribs → OA → the rest of the cervical spine → shoulder → arm → hand

Lower thoracic spine → lower ribs

L5 → the rest of the lumbar spine → psoas or related structures → leg → foot
          ↓
      sacrum

Clinical Note: Cranial treatment can be implemented prior to any of the above sequences in order to create a relaxed state and to augment OMT in other areas.

TREATMENT TYPES

Technique	Direct		Indirect	Active		Passive
Counterstrain			X			X
Facilitated Positional Release			X			X
Myofascial Release	X	or	X	X	or	X
Muscle Energy	X			X		
HVLA	X					X

Counterstrain
- developed by Lawrence Jones, DO
- a tender point is engaged and the body part is "folded" around the tender point to gain the maximum point of ease/comfort, and then is held in that position for a specific period of time (90 seconds or 120 seconds, depending upon body region)
- goal = to reduce tenderness by 70% or more
- if a tender point is non-responsive to Counterstrain technique, it is referred to as a Maverick point

Facilitated Positional Release (FPR)
- developed by Stanley Schiowitz, DO
- the body part is placed into a position that is neutral in all planes and that ensures the least joint and tissue tension in all 3 planes; then, an "activating force" (compression or torsion) is applied. After that, the body region is moved further away from the barrier and held for 3-4 seconds.
- is intended only for the treatment of superficial muscles and the deep intervertebral muscles

Myofascial Release
- first described by Andrew Taylor Still
- the myofascial tissues are engaged with a constant force guiding them towards the barrier (direct) or away from the barrier (indirect); this force is held for a specific period of time or until a release is felt
- some authorities dispute that myofascial release can utilize active forces; most texts on the subject do acknowledge, however, that an active or "activating force" can be used by the patient to increase effectiveness of the technique. Examples of such active forces include inhaling or exhaling during the technique according to the clinician's instructions and in

conjunction with the manipulation, holding one's breath at a certain point and in a certain way during the manipulation, rolling one's eyes upward, etc.

Muscle Energy
- developed by Fred S. Mitchell, DO
- this is not recommend for post-op or ICU patients (because it is an active, direct technique)
- the physician directs the body part, mindful of all three planes, into the barrier; the patient then is directed to actively move the body part in precisely the opposite fashion into which it has been placed, or in precisely the same direction into which it has been placed, depending upon which type of muscle energy is utilized. All the while, the physician maintains that body part in the original position into which he/she placed it (i.e., the body part is not permitted to move). This process is repeated 3-5 times.
- there are two types, depending upon which muscles are engaged during the technique:

    1) <u>Post-isometric relaxation</u>: the muscle causing the restriction is engaged and contracted (and, because no movement is allowed, the contraction is isometric). Activation of the Golgi tendon organs allows for reflex relaxation of the agonist muscle.

    2) <u>Reciprocal inhibition</u>: the muscles antagonistic to the muscle causing the restriction are engaged and contracted (and, again, no movement is permitted; but it is the reciprocal - or antagonist - muscles that are contracted). Activation of the reciprocal inhibition reflex arc in the spinal cord allows for reflex relaxation of the agonist muscle. *(Please note that occasionally, but rarely, this type of muscle energy is performed with an indirect set-up as opposed to the more typical direct approach.).*

High Velocity Low Amplitude (HVLA)
- developed by A.T. Still
- "thrust treatment" or "mobilization with impulse treatment"
- the restrictive barrier of the body part is engaged firmly and securely; then a high velocity (very fast), low amplitude force (minimal force) is applied to that body part/joint in the direction of the pathologic barrier.
- the thrust is always performed only during the relaxation phase (exhalation phase).

Chapter 1 Review Cases

1. A patient presents with complaints of back pain. On evaluation, you note that her T5 segment easily rotates to the right, but rotates only a little to the left. What is a true statement?
   a. The T5 left anatomic barrier can be reached
   b. The T5 complete limit on the right is the physiologic barrier
   c. The left physiologic barrier of T5 cannot be reached
   d. The right anatomic barrier of T5 cannot be reached
   e. A restrictive barrier is engaged when T5 is rotated right

2. A 33-year old male presents for evaluation. On examination of L4, you note that your left thumb goes in deeper than your right thumb. You note that this discrepancy is particularly increased when L4 is flexed; neutral positioning of L4 restores near-symmetric left versus right findings over the transverse processes. What is the dysfunction that exists?
   a. L4NSLRR
   b. L4NSRRL
   c. L4FRRSR
   d. L4FRLSL
   e. L4FRRSL

3. With respect to the previous patient, what is the restriction?
   a. L4ERLSR
   b. L4NRRSR
   c. L4FRLSR
   d. L4FRLSL
   e. L4NRLSR

4. A patient was hospitalized for an emergent cholecystectomy. You wish to perform OMT to relieve some of the patient's somatic complaints. Which of the following methods would be most appropriate for this patient?
   a. HVLA
   b. Reciprocal Inhibition
   c. Thrust treatment
   d. Post-isometric relaxation
   e. Myofascial release

5. A 45-year-old male presents with complaints regarding his right thigh. Where will you begin your evaluation and treatment process?
   a. sacral spine
   b. right thigh
   c. lumbar spine
   d. pelvis
   e. lower thoracic spine

6. On physical examination, a patient demonstrates T5 rotated right; this is most noticeable in flexion, and least noticeable in neutral. What is the patient's somatic dysfunction?
    a.   T5FRRSR
    b.   T5ERRSR
    c.   T5NSLRR
    d.   T5NSRRL
    e.   T5FRLSL

7. With respect to the previous patient, what is the diagnosis?
    a.   T5FRRSR
    b.   T5ERRSR
    c.   T5NSLRR
    d.   T5NSRRL
    e.   T5FRLSL

8. With respect to the previous patient, what is the restriction?
    a.   T5FRLSR
    b.   T5ERRSR
    c.   T5NSLRR
    d.   T5NSRRL
    e.   T5FRLSL

9. As a patient forward bends at the waist, in what plane and around what axis is he moving, respectively?
    a.   Transverse plane; vertical axis
    b.   Coronal plane; A-P axis
    c.   Sagittal plane; transverse axis
    d.   Frontal plane; A-P axis
    e.   A-P plane; A-P axis

10. A patient presents with a contracture of the right hamstrings. If you want to perform reciprocal inhibition to treat this contracture, into what position will you place the patient, and what movement will you ask him to isotonically undertake, respectively?
    a.   Flexion at the hip; flexion at the hip
    b.   Flexion at the hip; extension at the hip
    c.   Extension at the hip; extension at the hip
    d.   Extension at the hip; flexion at the hip
    e.   Abduction at the hip; adduction at the hip

Answers to Chapter 1 Cases

1. C   By easily rotating T5 all the way to the right, one is taking that segment to its limit in terms of right rotation; thus, the right anatomic barrier is reached. However, there is restriction in left rotation, preventing even the physiologic barrier from being reached. Hence, the barrier that one encounters on the left is the restrictive barrier.

2. A   If the left thumb goes in with greater ease, that means that the segment is rotated right. Thus, L4 is rotated right. However, in evaluating flexion versus extension versus neutral, one finds that the dysfunction is worsened by flexion; in other words, flexion removes the segment itself from freedom, locking it up further into a dysfunctional position. Conversely, neutral positioning is the position in the sagittal plane that restores near symmetrical left and right rotation, freeing that segment up to enjoy the most overall freedom. Thus, L4 is neutral (the position of most freedom). Given that it is neutral, sidebending and rotation must occur in opposite directions, per Fryette principles. Since we know it is rotated right, it must therefore be sidebent left. The dysfunction is L4NSLRR.

3. C   If its freedom (dysfunction) is L4NSLRR, then we know that its restriction must be the opposite of that position of most freedom. So, if the freedom is rotating right, then the restriction is rotating left. If the freedom is sidebending left, then the restriction is sidebending right. Lastly, as is explained in answer #2, flexion is that position that confers that greatest restriction ("locking up") of the segment. Thus, the restriction is L4FRLSR. Restriction does not follow Fryette mechanics; restriction is simply the opposite of the freedom.

4. E   Myofascial release is the only one of the techniques noted that can be generally performed as an indirect technique. Given that the patient is hospitalized, direct techniques (those techniques that place the patient directly into their restrictive barrier) are not recommended.

5. C   In evaluating and treating a patient, one must always begin centrally. Only by starting centrally can the cause eventually be identified and treated. The causes for many complaints are not necessarily always at the site of discomfort. Thus, all lower extremity complaints require that one begin one's assessment at the level of the lumbar spine.

6. C   According to the physical exam, T5 is rotated right. However, this rotation asymmetry is most noticeable in flexion. Remember, rotation in a healthy adult should be symmetric from side to side. So, the finding of increased rotation on one side implies that there is restriction on the other side; in other words, there is dysfunction. We name that dysfunction by the freedom of movement for that plane; for rotation, the greatest freedom is rotation right. The greater the rotation to one side, the greater the asymmetry; the greater the asymmetry, the greater the dysfunction. Hence, in this patient, as rotation to the right is maximized in flexion, flexion can be said to exacerbate the dysfunction. It is the position in the sagittal plane that further "locks" the spine up into dysfunction. So, flexion is the position in the sagittal plane that most locks the spine up. However, the physical exam also revealed that the rotational asymmetry was least noticeable in neutral. So, neutral allows for the least amount of asymmetry and the least amount of dysfunction. In other words, neutral position is that position that allows for the most freedom in the sagittal plane. Remember, we always name dysfunctions by the freedom for that particular plane or movement. The most freedom for ROTATION (transverse plane) was rotation right. The most freedom for the SAGITTAL PLANE is neutral. Thus, we know that T5 is neutral and rotated right; applying Fryette mechanics, we know that, since it is a neutral dysfunction, sidebending must be in the opposite direction. Thus, the somatic dysfunction is correctly noted as T5NSLRR.

7. C   The diagnosis is the somatic dysfunction, and is what is noted in the patient's chart.

8. A   Restriction is the OPPOSITE of freedom. Restriction does not follow Fryette mechanics. The only rule that restriction follows is that it has to be the opposite of the freedom in all three planes. Thus, if the dysfunction was T5NSLRR, we know that the freedom was the same: T5NSLRR. Thus, the restriction in the sagittal plane is either flexion or extension; however, based on the patient's original findings that flexion made the spine most locked up into a dysfunctional position, we know that the restriction is flexion. If the freedom is sidebending left, then we know that the restriction is sidebending right. If the freedom is rotating right, then the restriction is rotating left. Hence, the restriction is T5FSRRL; remember, restriction does not follow Fryette mechanics!

9. C   As the patient is forward being, he is moving through the sagittal plane (also known as an A-P plane) about a transverse axis.

10. A If there is a contracture of the hamstrings, that means that the hamstrings are contracted or tightened to induce extension of the leg at the thigh. Reciprocal inhibition is a type of muscle energy. Any muscle energy generally requires that the patient be placed into his restrictive barrier. Well, if the leg is, as a result to the dysfunction, being extended, then the freedom is extension. That means that the restriction is flexion at the hip. Thus, the set-up is to place the patient's thigh into flexion as much as possible, until the restrictive barrier is met. Reciprocal inhibition requires that the muscles reciprocal to or antagonistic to the muscles in dysfunction be engaged into contraction, thereby forcing the agonist muscles (in this case, the dysfunctional muscles) to relax. If the hamstrings are in dysfunction, and their normal function is to cause extension of the thigh at the hip, then the antagonist muscles would be the quadriceps since they cause the opposite movement, specifically flexion at the hip. Thus, after the patient's leg is placed into its restrictive barrier by taking it into a flexive position as much as possible, the patient is then asked to flex his leg at the hip (while you hold the leg still so as to allow for an isotonic contraction).

Chapter Two: CERVICAL SPINE

Bones
The typical cervical vertebra (C3-C6):

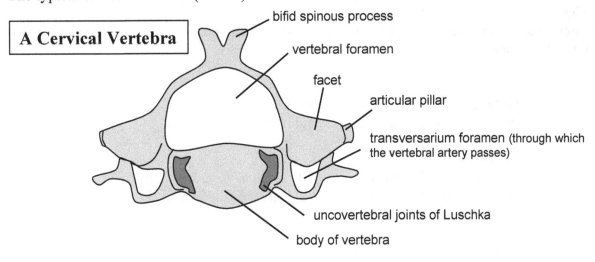

| A Cervical Vertebra |

- bifid spinous process
- vertebral foramen
- facet
- articular pillar
- transversarium foramen (through which the vertebral artery passes)
- uncovertebral joints of Luschka
- body of vertebra

Cl, C2, and C7 are not as pictured above.

C1 **(atlas)** and C2 **(axis)** are atypical. Cl has no spinous process or vertebral body. C2 has an **odontoid process** (<u>dens</u>).

C7 is not considered *atypical* per se, but it is different from the other cervical vertebrae in that it does not have a bifid spinous process. Its spinous process is not divided, but is very long; its spinous process is termed the **vertebra prominens.**

The mid-cervical vertebrae have an unusual set of synovial joints found on the lateral surface of their bodies. These joints are known as the **uncinate** or **uncovertebral joints of Luschka.** Their function is to provide stability to that region of the spine and to lessen the possibility of a herniated nucleus pulposus at each level. The other joints that exist are the **zygopophyseal joints** (<u>facet joints</u>) which exist between the superior facet of the lower vertebra and the inferior facet of the upper vertebra.

Ligaments

Alar ligament:	connects the sides of the odontoid process to the lateral aspects of the foramen magnum.
Transverse ligament:	extends between the two C1 lateral masses and holds the dens in place; it is posterior to the dens
Cruciform ligament:	comprised of 4 components - the transverse ligament (described above), the superior band (passes from the transverse ligament to the occiput), and the inferior band (passes from the transverse ligament to the body of C2)

Tectorial Membrane: the superior continuation of the posterior longitudinal ligament; lies just posterior to the cruciform ligament and passes from C1 to the internal surface of the occipital bone

Anterior Longitudinal Ligament: runs anteriorly over the spine

Posterior Longitudinal Ligament: runs over the posterior aspects of the vertebral bodies, anterior to the spinal cord

Muscles

Scalenes (anterior, middle, and posterior) - run from the lateral portions of the cervical spine to the 1st and 2nd ribs

Unilateral contraction: sidebend the neck
Bilateral contraction: flex the neck; aid in forced respiration

respiratory assistance:
ANTERIOR & MIDDLE SCALENES - aid in elevation of the first rib
POSTERIOR SCALENE - aids in elevation of the second rib

Clinical Note: A tender point in a scalene may denote a 1st or 2nd inhalation rib dysfunction.

Trapezius: the primary connection between the neck and the head and shoulder girdle.

Sternocleidomastoid: passes from the mastoid process to the sternum and clavicle.

Nerves

8 Cervical Nerve Roots:
Cervical nerve roots C1-C7 emerge above their corresponding vertebra: C8 emerges below cervical vertebra 7.

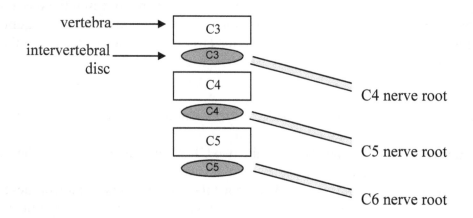

Brachial Plexus: originates from C5-T1 nerve roots.

<u>Motion and Function</u>
Fryette principles DO NOT apply to the cervical spine!

CERVICAL MOTION RULES

Segment	Primary Motion	Rot & SB	Evaluation via
Occipitoatlantal (OA)	F/E	opposite	Translation
Atlantoaxial (AA)	Rot	opposite	Rotation
Upper cervical	Rot	same	Translation
Lower cervical	SB	same	Translation

F/E = flexion/extension   Rot = rotation   SB = sidebending
OA = C0 = occiput = segment of the occipitoatlantal joint
AA = C1 = atlas = segment of the atlantoaxial joint

<u>Evaluating the OA⟶ Check Translation</u>
1. Induce translation (side to side movement); do in neutral and flexion/extension.
2. Right translation causes left sidebending:

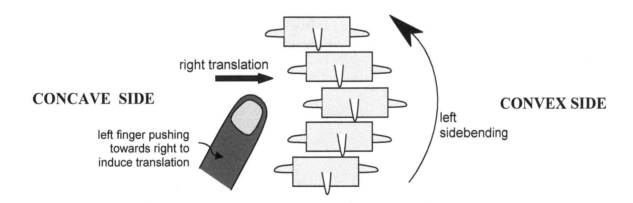

TRANSLATION AND SIDEBENDING

3. If there is restriction, that means that the translation is restricted. This means that there is restriction in sidebending (because, after all, translation induces sidebending)

As before, when indicating the sidebending dysfunction, always note the freedom of movement in one's nomenclature. And, observing the cervical rules of motion, we can assume that the rotation is in the opposite direction, noting it accordingly.

Example: If there is restriction of left OA translation, then the freedom of movement is with right translation. Therefore, the occiput is sidebent left. If it is sidebent left, by following cervical motion rules we know that it is rotated right. We call this (different nomenclature than elsewhere!) a posterior occiput right. (For simplicity we have not included the F/E vs. neutral component, but that would be noted too in terms of the position of freedom of movement)

Occiput = CO

Evaluating AA ⟶ Check Rotation
1.   Place one's fingertips on the lateral masses of C1
2.   Flex the neck to 45 degrees - to lock out C2-C7!
3.   Rotate to each side and determine any restriction in rotation

NOTE: Flexion, extension, and sidebending motions are not generally tested, but can be a source for written examination questions.

Example: If the atlas is rotated left (with right rotation restriction), it is often referred to as a posterior atlas left or an anterior atlas right.

Clinical note: Atlas dysfunction frequently causes retro-orbital pain.

Evaluating C2-C7 ⟶ Check Translation
1. Place one's fingertips on the lateral portions of the anterior cervical pillars.
2. Induce translation (and, therefore, segmental sidebending) - note any restriction in neutral, flexed, and extended positions. Notation of the dysfunction is performed in the standard format for spinal somatic dysfunction.

Articular Pillars = Cervical Pillars

Clinical Note on Treatment of the Cervical Spine: Generally, whenever evaluating and treating the cervical spine, it is preferred to begin with the OA and then treat the rest of the cervical spine.

Chapter 2 Review Cases

1. If a patient's transverse ligament of the cruciform ligament is ruptured, what structure might the dens penetrate?
   a. Brainstem
   b. Spinal cord
   c. Nerve root
   d. Vertebral artery
   e. Intervertebral disc

2. A 32 year-old patient is determined to have a herniated C5 intervertebral disc. What nerve root risks impingement?
   a. None
   b. C3
   c. C4
   d. C5
   e. C6

3. While evaluating a patient, you note that the occiput is rotated right and that this rotation is most evident in extension. Which of the following could be a correct diagnosis?
   a. C0ERRSL
   b. OAFRRSL
   c. C0FRRSR
   d. OAERRSR
   e. C1FRRSL

4. With respect to the previous patient, what is the restriction?
   a. C0ERLSL
   b. OAERLSR
   c. C0FRLSL
   d. OAERRSR
   e. C1FRLSL

5. You are in the process of doing a full spinal assessment of a patient. You are about to start actually evaluating and, if necessary, manipulating the cervical spine itself. Where should you start?
   a. OA
   b. AA
   c. Lower cervicals
   d. C7
   e. T1

6. A patient presents with complaints of pain in his left neck. His head appears to be sidebent to the left. Physical exam reveals a spasm of the left middle and anterior scalenes. What other somatic dysfunction do you expect to find?
   a. 1st inhalation rib
   b. 1st exhalation rib
   c. 2nd inhalation rib
   d. 2nd exhalation rib
   e. 3rd-5th inhalation rib

7. If C3 is rotated right and neutral, what is the restriction?
   a. C3NRRSL
   b. C3NRRSR
   c. C3FRLSL
   d. C3FRLSR
   e. C3ERRSR

8. What anatomic structure passes through the transversarium foramen of the cervical vertebrae?
   a. Spinal cord
   b. Brain stem
   c. Nerve roots
   d. Cervical sympathetic chain ganglia
   e. Vertebral artery

9. A patient presents with complaints of paresthesias in an area of their arm that you know to represent the C8 dermatome. If this is due to a herniated intervertebral disc, what disc is responsible?
   a. C5
   b. C6
   c. C7
   d. C8
   e. T1

10. On examination, you find the OA rotated to the right, with least rotational asymmetry is noted in flexion. What is the somatic dysfunction that exists?
    a. OANRRSL
    b. OcciputFRRSR
    c. AtlasNRRSL
    d. AAFRRSL
    e. C0FRRSL

11. What would be another correct way to describe this dysfunction?
    a. Posterior occiput right
    b. Anterior occiput left
    c. Posterior atlas right
    d. Anterior atlas left
    e. Posterior axis right

12. As you are palpating a patient's spine, you note that you are more easily able to move the spine to the right with your left thumb than to move it to the left with your right thumb. What motion are you able to more easily induce?
    a. Left translation
    b. Right translation
    c. Flexion
    d. Extension
    e. Subluxation

13. With respect to the previous patient, what is the freedom of movement, and to what side is a convexity most easily created, respectively?
    a. Right sidebending, right
    b. Left sidebending, right
    c. Right sidebending, left
    d. Left sidebending, left
    e. Flexion and Extension, left

Answers to Chapter 2 Cases

1. B   The brainstem is protected within the skull. The dens is part of C2, and this is part of the spine. The spine protects the spinal cord. So, a ruptured transverse ligament could allow the dens to deviate posteriorly to penetrate the spinal cord. It should be noted, however, that the medulla, the lower most part of the brainstem, is the superior continuation of the spinal cord. If the dens is thrust upward, such as with hanging, it can have the potential to rupture the medulla.

2. E   Remember, in the cervical spine, the nerve roots emerge above the same-number vertebra. Also, throughout the spine, the disc sits beneath its own vertebra. So, if C5 disc is herniated, that means that the disc that is beneath C5 vertebra but above C6 vertebra is herniated. The C6 nerve root emerges above the C6 vertebra. Thus, it is the C6 nerve root that risks impingement.

3. B   If the occiput rotates more easily to the right, we know it is rotated right. The next step is to identify the freedom of movement in the sagittal plane: flexion, extension, or neutral. Thus, which of those positions affords the spine the most freedom or, in other words, the least amount of dysfunction? Well, if the occiput rotates more to the right rather than symmetrically while in extension, that is evidence that extension places the spine into more of a locked up or dysfunctional position. Hence, extension is the restriction. The freedom, thus, has to be either flexion or neutral. Lastly, to determine the sidebending component, we need only to consider what cervical unit this is. Remember, the cervical spine does not follow Fryette mechanics; it follows Cervical Motion Rules, instead. Therefore, the rotation and sidebending relationship is determined exclusively by what cervical segment is involved. Herein, the occiput is involved. The occiput is also known as the OA or C0. Rotation and sidebending always occur in opposite directions, regardless of the sagittal plane freedom. Hence, if the occiput is rotated right, we know that it has to be sidebent left. Therefore, the diagnosis is OA (or Occiput or C0), Flexed or Neutral, RRSL.

4. B   The restriction is the opposite of the freedom. The freedom is the diagnosis or somatic dysfunction. Thus, if the freedom (diagnosis), as noted in the previous question, is OAFRRSL, the restriction has to be the opposite: OA (or Occiput or C0), Extended (because that is the position, as noted in the previous question, that induces most dysfunction by way of inducing the most asymmetrical rotation), RLSR. Thus, of the options presented, the only correct one is OAERLSR. Remember, restriction found anywhere in the spine never follows Fryette mechanics; restriction is just the opposite of the freedom.

5. A   Of course, one should start centrally whenever evaluating any somatic complaint. Starting centrally, rather than at the site of pain or discomfort, increases the likelihood for identification and subsequent resolve of the actual cause of the pain or discomfort, since the source of the pain or discomfort may not be at the site of the pain or discomfort. Accordingly, if a patient presented with a cervical complaint, it is recommended that the evaluation start at the upper thoracic spine. However, this case notes that you are already about to start actually evaluating and manipulating the cervical spine itself. When one does start actually evaluating the cervical spine itself, the OA is the recommended site for initiation of that cervical spine evaluation.

6. A   Muscle spasms cause contractions that result in muscle shortening. The anterior and middle scalenes are attached to the first rib. Hence, if they are shortened, they will lift the rib up. If a rib is "up," its freedom is up. So, we name everything for its freedom. The proper term for ribs that are up is "inhalation," since that is the general position of a rib during the normal physiologic process of inhalation. Accordingly, in this case, even without inhalation, the freedom mimics the position normally seen during inhalation. So, we call this an inhalation rib and, since it involves the first rib, it is a first inhalation rib.

7. C   If C3 is rotated right, we know that the sidebending component has to be in the same direction, regardless of its free sagittal plane (e.g., neutral in this case). After all, cervical segments follow cervical motion rules, and C3 rotation and sidebending always go in the same direction. Hence, we know that the dysfunction or freedom is C3NRRSR. The question asks what the restriction is. Restriction is simply the opposite of the freedom. Thus, the restriction is C3 F/E RLSL.

8. E   It is the vertebral artery that passes through the transversarium foramen, and is the reason we must avoid doing cervical HVLA in those with known atherosclerotic disease.

9. C   The C8 dermatome, by definition, is that dermatomal region served by the C8 nerve root. The C8 nerve root, like any cervical nerve root, emerges from the spine at the level above the same-numbered vertebra; hence, it emerges above what is equivalent to the "C8 vertebrae:" T1 (as there is NO C8 vertebra). So, it emerges above T1 segment. What is above T1? The C7 vertebral segment sitting upon its disc, the C7 disc. Thus, it is the C7 intervertebral disc (also known as the C7-T1 disc) that is the culprit in causing C8 nerve root impingement.

10. E   If OA is rotated right, you know, from cervical motion rules, that it must also be sidebent left (e.g., to the opposite side). The least rotational asymmetry is noted in flexion. Well, freedom in the sagittal plane (e.g., flexion, extension, neutral) is always gained when there is symmetry. So, freedom in the sagittal plane in this case is gained in flexion. Thus, the dysfunction is properly recorded as OA F RRSL. Since the OA is also known as C0, this can also be recorded as C0 FRRSL.

11. A   This type of nomenclature names what side of that anatomic part protrudes the most. So, a posterior occiput right implies that the right posterior occiput protrudes more than the left posterior occiput. That would be consistent with an occiput that is rotated right, such as in this case.

12. B   If you move the spine laterally to the right with your left thumb, you are causing it to translate (change positions) to the right.

13. B   If you are able to translate it more to the right, that means that the freedom is that of translation to the right. Well, if you induce several vertebrae to move to the right, you will cause that area of the spine to "stick" out in the area to the right – that causes a right convexity. But, it is only misaligned to the right in that one area, with the rest of the spine remaining in its original position. Therefore, overall, you are inducing one part of the spine to move to the right, while the rest of it stays "more left" of that rightward area; that yield left sidebending.

Chapter Three: THORAX

Structure
The bony thorax is comprised of the thoracic spine, the ribs, and the sternum.

Anatomy of the Thoracic Spine
- **12 thoracic vertebrae**
- each vertebra has a very long spinous process which has varying degrees of curvature, depending upon the vertebra of origin. Refer to the "Rule of Threes" for more detail.

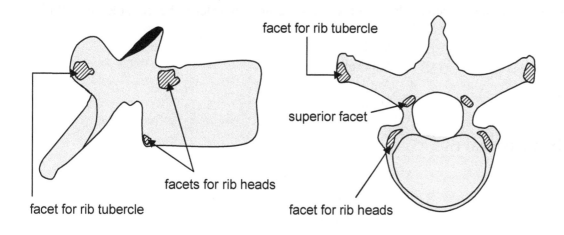

facet for rib tubercle

superior facet

facet for rib heads

facets for rib heads

facet for rib tubercle

Thoracic Vertebra

The facets for rib heads and rib tubercles are collectively known as **costal facets**. Superior and inferior facets are collectively known as **articular facets**, as they are involved with articulation with superior and inferior vertebrae, respectively. Articular facets are sometimes termed **zygopophyseal facets**.

A facet for a rib tubercle articulates with the rib tubercle of the same-numbered rib. The superior facet for the rib head also articulates with the same-numbered rib, doing so at the rib's head. The inferior facet for the rib head articulates with the rib head of the next-numbered rib (e.g., the head of rib 11 articulates with the inferior facet for the rib head found on T10). Thus, a rib head actually articulates with 2 vertebrae: the same-numbered vertebra and the one-less-numbered vertebra. So, for example, the head of rib 11 articulates with T11 (at the superior facet for the rib head of T11) and with T10 (at the inferior facet for the rib head of T10).

Rule of Threes:
T1-T3→ the transverse process is located on the same horizontal plane as the spinous process of that given vertebra.

T4-T6→ the transverse process of any given vertebra is located halfway between that vertebra's spinous process and the spinous process of the vertebrae above it.

T7-9 →the transverse process of any given vertebra is located on a horizontal plane corresponding with the spinous process of the vertebra directly above it.

T10-12→each of these act uniquely with respect to each other.

> T10 obeys the same rule as that for T7-T9.
> T11 obeys the same rule as that for T4-T6.
> T12 obeys the same rule as that for T1-T3.

Anatomy of the Ribs
There are **12 ribs.** They are classified as either typical or atypical ribs.

Typical Ribs - must have each of the 5 following structures in order to be considered as typical.

> 1. Head
> 2. Neck
> 3. Tubercle
> 4. Angle
> 5. Shaft

Ribs 3-10 are typical ribs

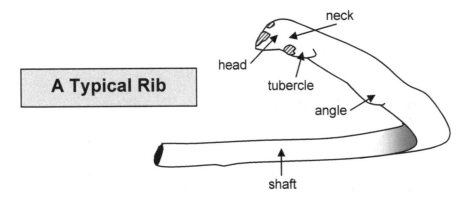

Atypical Ribs - these are the ribs that do not have all five anatomical structures (or have more than these) required to be considered as typical. Ribs 1, 2, 11, and 12 are atypical ribs.

Rib 1: is atypical because it has no angle; also, although it has a head, that head only articulates with the corresponding vertebra (T1), failing to articulate with two different vertebrae as all typical ribs do.

Rib 2: is atypical because it has an extra tuberosity on its shaft.

Ribs 11 & 12: they are atypical because they have no neck or tubercle; also, although they each have a head, that head only articulates with the corresponding vertebra, failing to articulate with two different vertebrae as all typical ribs do.

> *Note:* Occasionally, rib 10 is also considered atypical since its head also only articulates with its corresponding vertebra.

The broadest and most curved rib = rib 1

Ribs can also be classified as **true, false,** or **floating ribs.**

True Ribs: ribs that are connected to the sternum by their costal cartilage. Ribs 1-7 are true ribs.

False Ribs: ribs that are not DIRECTLY connected by their costal cartilage to the sternum. Ribs 8-10 are false ribs (they connect by their costal cartilage to the rib superior to them).
Ribs 11-12, although more usually considered only as "floating ribs," are technically also false ribs (they do not connect to the sternum at all)

Floating Ribs: ribs that do not connect to the sternum at all; a type of false rib. Ribs 11-12 are floating ribs.

FUNCTION
The functions are motion, support, and protection.

Motion of the Thoracic Spine
The primary motion of the thoracic spine is *ROTATION*

T11-12 gain motion parameters more akin to lumbar vertebrae.

Hence, the order of ease for the thoracic spine motion is as follows:
rotation>flexion/extension>sidebending

The order of ease for T11-12 motion is as follows:
flexion/extension>sidebending>rotation

Motion of the Ribs
The three types of rib motion:
1. **Pump-handle**
2. **Bucket-handle**
3. **Caliper**

The type of rib motion a particular rib most (MOST, not only) undertakes is dependent on exactly which rib that is:
Ribs 1-5 (upper ribs): pump-handle motion
Ribs 6-10 (middle ribs): bucket-handle motion
Ribs 11-12 (lower ribs): caliper motion

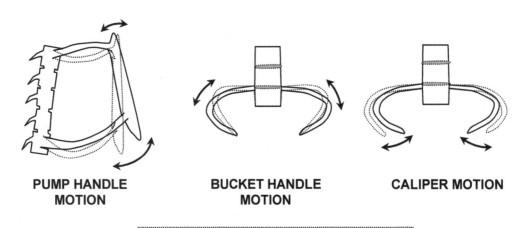

PUMP HANDLE MOTION   **BUCKET HANDLE MOTION**   **CALIPER MOTION**

Types of Rib Motion

<u>Muscles of respiration</u>
Primary muscles:
> DIAPHRAGM
> INTERCOSTALS (external, internal, innermost, and subcostal)

Secondary muscles:
> SCALENES
> PECTORALIS MINOR
> LATISSIMUS DORSI
> QUADRATUS LUMBORUM
> SERRATUS ANTERIOR and POSTERIOR

Diaphragm
1. contracts with inspiration, relaxes with exhalation
2. the contraction/relaxation causes a pump-like action to aid in the return of lymph and venous blood back to the thorax.
3. attachments = ribs 6-12 bilaterally, bodies and intervertebral discs of L1-L3, and the xiphoid process.
4. innervation = phrenic nerve (originates from C3-5 nerve roots)

Intercostals
1. lift ribs with inhalation, and prevent retractions during inspiration

RIB DYSFUNCTION

Two types of rib dysfunctions:

1. **Inhalation dysfunction** = exhalation restriction (antiquated term)
   Remember, we always name dysfunctions for their ease of motion. In this case, the rib is

in a position it would engage as if it were engaging in inhalation; for instance, for a rib that undertakes bucket handle motion, it would be up as in inhalation, meaning it is restricted in coming down (as in exhalation). So, its freedom of movement is actually up. Hence, it is called an Inhalation Dysfunction.

2. **Exhalation dysfunction** = inhalation restriction (antiquated term)
   The ribs are "stuck" down or in a position similar to that they would undertake physiologically during exhalation (with the difference being that they stay in the exhalation-like position)

Group Rib Dysfunction
   ➢ a group of dysfunctional ribs
   ➢ usually due to just one rib that is causing dysfunction in the others. The responsible rib is the **key rib.**

KEY RIBS
For Inhalation Dysfunction: the lowest rib of the group.
For Exhalation Dysfunction: the highest rib of the group

Important Anatomical Landmarks
Sternal notch is at the same level as T2.
Spine of the scapula is at the same level as T3.
Sternal angle (Angle of Louis) is at the same level as T4.
Inferior angle of the scapula is the same level as T7.

Chapter 3 Review Cases

1. A patient is determined to have a congenital deformity of the facet for the rib tubercle on the T5 segment. What type of facet is deranged?
    a. Articular facet
    b. Costal facet
    c. Dens facet
    d. Zygopophyseal facet
    e. Multifacet

2. While palpating the spine of a 69 year old Caucasian female, you are assessing each vertebra, one by one as you go down the spinal column. When you reach the tip of spinous process of T8 and move your thumbs directly laterally, you find that the underlying segment's transverse processes are deeper on the right than on the left. You also determine that this is a Type I Fryette dysfunction. What is the most accurate way to record this dysfunction?
    a. T8FRLSL
    b. T8NSRRL
    c. T8ERRSR
    d. T9FRLSL
    e. T9NSRRL

3. A 32-year male suffers severe injuries as the result to a motorcycle accident. His T7 and T8 vertebral segments are both shattered. Which of the following rib will have no vertebral articulation as a result to this injury?
    a. Rib 5
    b. Rib 6
    c. Rib 7
    d. Rib 8
    e. Rib 9

4. With respect to the previous patient, which of the following is a correct statement?
    a. One of his true ribs has no vertebral articulation
    b. One of his false ribs has no vertebral articulation
    c. One of his floating ribs has no vertebral articulation
    d. One true rib and one false rib each have articulation with only one vertebra, while one false rib has no articulation with any vertebrae
    e. One true rib and one false rib each have articulation with only one vertebrae, while one true rib has no articulation with any vertebrae

5. In palpating a patient's ribs, you find that the angle of the rib is more superior than either the rib head or its attachments anteriorly. What movement is it undertaking?
   a. Pump handle
   b. Love handle
   c. Bucket handle
   d. Door handle
   e. Caliper

6. A patient presents with complaints of shortness of breath. During the physical examination, you note that there are a number of other somatic dysfunctions, as listed below. Which of the following may be responsible for her shortness of breath?
   a. Cervical somatic dysfunction of cervical segments C2-C6
   b. Anterior innominate rotation
   c. Hypertonicity of the gastrocnemius
   d. Backward sacral torsion
   e. Symmetric rotation to the right and left of thoracic vertebral segments T2-5

7. While evaluating a 73 year old female patient, you find that the head of rib 4 is significantly lower than the anterior tip of the rib. What dysfunction exists?
   a. Inhalation restriction of rib 4
   b. Exhalation dysfunction of rib 4
   c. Inhalation dysfunction of rib 4
   d. No dysfunction of rib 4 exists
   e. Group dysfunction of rib 4

8. At a later date, the aforenoted patient is again evaluated. On this later examination, ribs 3 and 4 now both exhibit the same previous anomaly. Which one is the key rib, and to which rib will you direct treatment, respectively?
   a. Rib 3, rib 3
   b. Rib 3, rib 4
   c. Rib 4, rib 3
   d. Rib 4, rib 4
   e. Rib 5, rib 5

9. Rib 11 is in exhalation restriction. What position would be the predominant one appreciated on palpation of this rib?
   a. The rib angle is positioned superiorly to both the anterior tip of the rib and head of the rib
   b. The anterior tip is superior to the rib head and angle
   c. The anterior tip is depressed compared to its normal position
   d. The anterior tip of the rib is more lateral than normal
   e. The anterior tip of the rib is more medial than normal

Answers to Chapter 3 Cases

1. B   Any facet that articulates with a rib is classified as a costal facet.

2. E   Since your right thumb goes in deeper than your left over the transverse processes, this is a segment that is rotated left. Since it states that it is a Fryette 1 dysfunction, you also know that this segment is in Neutral. Because it is neutral, the sidebending component must be opposite of the rotational component; thus it is sidebent right. Thus, the segment is NSRRL; but what segment is in this dysfunction? T9. According to the rules of three, the tip of the spinous process of T8 is so long that it is on the same horizontal level as the transverse processes of T9. Accordingly, if one moves one's thumbs directly laterally from the tip of the T8 spinous process, that will place the thumbs squarely over the transverse processes of T9. Hence, it is T9 that one was palpating. Therefore, the dysfunction is correctly noted as T9NSRRL.

3. D   A thoracic vertebra articulates with two ribs: At its superior facet for the rib head and at its facet for rib tubercle it articulates with the same-numbered rib, and at its inferior facet for the rib head it articulates with the next-numbered rib. Hence, T7 segment articulates with ribs 7 and 8. T8 vertebral segment articulates with ribs 8 and 9. It is rib 8 that will lose all vertebral articulation since its head normally articulates with vertebral segments T7 and T8, and its tubercle articulates with T8.

4. D   While it was established in the previous question that rib 8 has no articulation with a vertebra, there are other ribs partially affected. Rib 7 normally articulates with T6 and T7, and so currently retains articulation with only one vertebra since the T7 segment was shattered. Likewise, rib 9 normally articulates with T8 and T9, and so currently retains articulation with only one vertebra since the T8 segment was shattered. Thus, while rib 8 retains no vertebral articulation, ribs 7 and 9 retain only single-segment articulation. Rib 7 is a true rib, and ribs 8 and 9 are false ribs. Thus, as a result of this injury, one true rib (rib 7) articulates with only one vertebra (T6), one false rib (rib 9) articulates with only one vertebra (T9), and one false rib (rib 8) articulates with no vertebrae (since T7 and T8 were shattered).

5. C   If the front and back of the rib are lower than an area more middle of the rib, we can see that the rib is acting like a bucket handle. So, this is bucket handle movement.

6. A   Shortness of breath can be caused by many things. However, the diaphragm must always be considered as a potential culprit in any dyspnea cases. As such, any attachments or innervations to the diaphragm should also be considered. The innervation to the diaphragm is the phrenic nerve, which is derived from cervical nerve roots C3 through C5. Cervical somatic dysfunction, including that of the vertebral segments, can, of course, impair normal function of those nerve roots. As a result, cervical somatic dysfunction of cervical segments C2-C6, as noted in the case, could impair adequate innervation to the diaphragm, resulting in shortness of breath.

7. C   Rib 4 undergoes primarily pump handle motion. During inhalation and anytime the rib is "stuck" in an inhalation-like position, the anterior part of the rib will be superior to the posterior part of the rib (the rib head), such as is found in this case. If the rib is "stuck" in such a position, it means it cannot go down. So, its "freedom" is that of being stuck up, and being restricted from going down. Therefore, this is an inhalation dysfunction, otherwise known as an exhalation restriction, of rib 4.

8. D   Now there is a group dysfunction (e.g., more than one rib collectively suffering from the same dysfunction). Both are in an inhalation dysfunction. The key rib is the responsible rib. In any inhalation dysfunction, the responsible rib is the bottom-most rib, as that is the rib that somewhat holds or pushes the ribs above it up. Thus, the key rib is rib 4. Likewise, the responsible rib, otherwise known as the key rib, is the rib to which treatment should be directed. Thus, treatment should also be directed towards rib 4.

9. D   If a rib is in exhalation restriction, it means it is in inhalation dysfunction. Thus, its freedom, or the position it most easily moves into, is that of inhalation. Rib 11 undergoes primarily caliper motion. Thus, during an inhalation-like position, it is "opened up," like a caliper. Accordingly, the anterior tip of the rib is more lateral than normal.

Handbook of OMT Review

Chapter Four: LUMBAR SPINE

Important Concepts:

The lumbar vertebrae have vertebral bodies that are designed to **carry vertical loads**.

The **posterior longitudinal ligament begins to narrow** in the lumbar spine. At L4-L5, that ligament is only <u>half</u> as wide as what it is at Ll. This creates a weakness in the posteriolateral aspects of the bodies and their corresponding intervertebral discs. Because of this lack of posteriolateral support in combination with the great vertical forces placed on the lumbar spine, this area of the spine is **particularly susceptible to disc herniation**.

When a herniation occurs, it affects the nerve root the level (or more) below the segment affected. This is because the nerve root for any given segment emerges from the intervertebral foramen above its corresponding intervertebral disc.

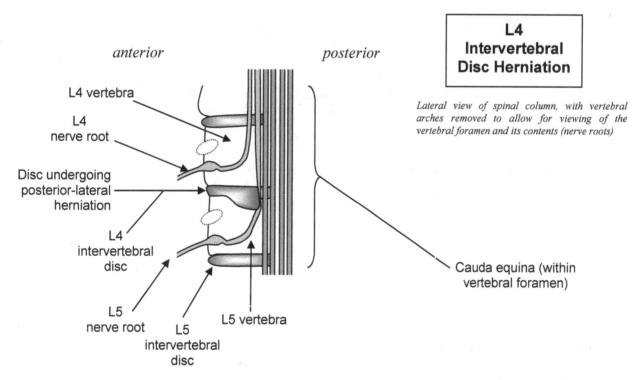

anterior   *posterior*

L4 vertebra

L4 nerve root

Disc undergoing posterior-lateral herniation

L4 intervertebral disc

L5 nerve root

L5 intervertebral disc

L5 vertebra

L4 Intervertebral Disc Herniation

Lateral view of spinal column, with vertebral arches removed to allow for viewing of the vertebral foramen and its contents (nerve roots)

Cauda equina (within vertebral foramen)

The spinal cord ends at L1-2 in most people. This is because the spinal cord segments do not align with the vertebral segments in the thoracic spine. Hence, the nerve roots for the lumbar spine must extend all the way from the upper lumbar spine to their final site of exit. Therefore, that nervous tissue in the vertebral canal inferior to the end of the spinal cord is termed the **cauda equina** because the *many* nerve roots mimic the appearance of a horse tail (cauda equina means horse tail).

Thus, the **cauda equina** is typically located distal to L2.

Typical site of lumbar puncture = L4-5

Muscles of the Lumbar Spine
1.   Multifidus
2.   Rotatores - the deepest of the muscles
3.   Erector Spinae (spinalis, longissimus, iliocostalis) - the mass of muscle that forms the
     prominent bulge on each side of the vertebral column
4.   Quadratus lumborum
5.   Iliopsoas (the iliacus muscles + the psoas muscle) - is an important muscle in
     maintaining the lumbosacral angle.
                    Iliacus = the primary flexor of the hip

Lumbosacral angle (Ferguson's angle)
Lumbosacral angle - formed by the intersection of a line representing the inclination of the sacrum
(this line passes through the PSIS) with a horizontal line that is in the same plane as the ASIS.

NORMAL LUMBOSACRAL ANGLE = 25 - 35 degrees
          (increased angle places increased stress on the lumbosacral joint, producing low back pain)

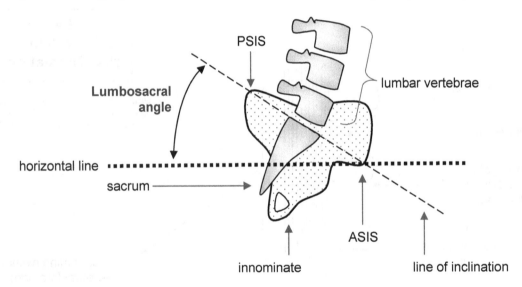

Lumbar Anatomic Anomalies

Sacralization : one or both of the transverse processes of L5 articulate with the sacrum.
Lumbarization: SI fails to fuse to the rest of the sacrum, instead remaining as a separate
                    vertebral segment.
Spina Bifida: failure or defect in the closing of the laminae.
                    laminae (of vertebral arch) = two bony plates that point posteriomedially
                    from the vertebral pedicles; their fusion forms the vertebral arch, the
                    posterior bony border of the vertebral canal. At their point of fusion, the
                    spinous process is directed posteriorly.

There are several types of spina bifida:

1.   SPINA BIFIDA OCCULTA - no herniation of spinal contents through the defect; often marked by a hairy patch or nevus on the skin over the site.

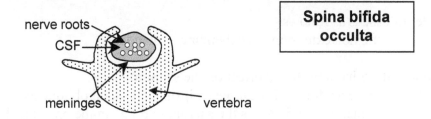

2.   SPINA BIFIDA MENINGOCELE - herniation of the meninges through the defect.

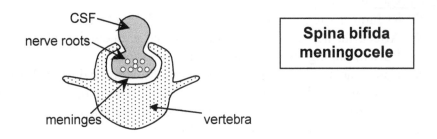

3.   SPINA BIFIDA MENINGOMYEOLOCELE - herniation of the meninges and nerve roots through the defect.

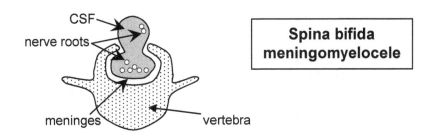

4.   RACHISCHISIS - a completely open spine

Any type of spina bifida characterized by a protruding sac is also categorized as a SPINA BIFIDA CYSTICA.

Also, be aware that spina bifida is not confined just to the lumbar region. It is most common in the sacral, lumbar, and lower thoracic regions, and usually extends for three to six vertebral segments.

Lumbar Spinal Motion

The primary motion of the lumbar spine is flexion and extension.

The order of ease is as follows:

flexion/extension > sidebending> rotation

The motion of L5 has a profound effect on the sacrum:
1.   Sidebending of L5 causes an ipsilateral sacral oblique axis.
2.   Rotation of L5 causes the sacrum to rotate in the opposite direction.

Lumbar Somatic Dysfunction

Herniated Nucleus Pulposus
- ❖ #1 lumbar site is between L4 and L5 or L5 and S1.
- ❖ Affects the nerve root below the herniated segment, i.e. a herniation between L4 and L5 will impinge upon L5 nerve root.
- ❖ Causes acute onset sharp, intense pain. May occur during flexion and/or carrying heavy weight. Pain is later exacerbated by flexion or carrying weight, and usually is described again as sharp. Often also causes sciatica.
- ❖ Can cause lower motor neuron pathology

Cauda Equina Syndrome
- ❖ Usually due to a massive central disc herniation resulting in high pressure on the nerve roots of the cauda equina
- ❖ Associated with same features as herniated nucleus pulposus, but also with saddle anesthesia in combination with urinary and/or fecal incontinence

Spondylolysis
- ❖ Fracture, disintegration, or dissolution of a vertebra
- ❖ Causes achy low back pain exacerbated by any activity
- ❖ A pars interarticularis fracture is usually seen radiographically, and is classically referred to as the *"collar"* on the Scotty dog on oblique views.

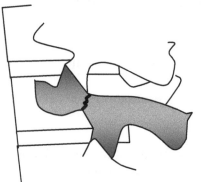

Oblique image of vertebra
Darkened part represents the "Scotty dog," complete with a fracture at the site of the pars interarticularis. This is a collar on the Scotty dog.

Spondylolisthesis
* *The anterior displacement of one vertebra with respect to the vertebra inferior to it*

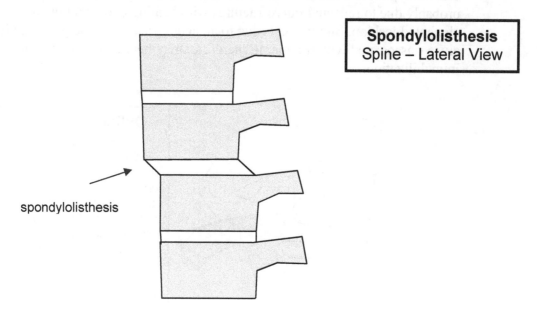

spondylolisthesis

* **There are 4 grades of spondylolisthesis** per the Meyerding's classification system which is based upon how much anterior displacement there is:
  Grade 1 = 25% anterior displacement
  Grade 2 = 50% anterior displacement
  Grade 3 = 75% anterior displacement
  Grade 4 = 100% anterior displacement

* Usually occurs between L4 and L5; is usually (not always) due to a fracture of the pars interarticularis (radiographically, demonstrates *a "beheaded Scotty dog"* on oblique views).
* Results in achy low back pain that is exacerbated by any activity; the acute event may or may not be associated with acute pain
* There are **five different classifications of spondylolisthesis**, depending upon the etiology of the spondylolisthesis:

Type I   Dysplastic Spondylolisthesis   *Congenital defect* involving deficiency of L5 or sacral neural arch and deficiencies of superior sacral facets; the L5 lumbosacral facets often are nearly horizontal. It is more common in girls.

Type II   Isthmic Spondylolisthesis   The most common type of spondylolisthesis and always involves *pathology of the pars interarticularis*. Subtype A is due to a fatigue fracture of the pars, and is the most common type of spondylolisthesis in people less than 50 years old. Subtype B is characterized by an intact but elongated pars, and is probably due to repeated microfractures with healing occurring in an elongated fashion. Subtype C is due to an acute fracture of the pars, and is always associated with trauma; it, therefore, may heal completely with immobilization.

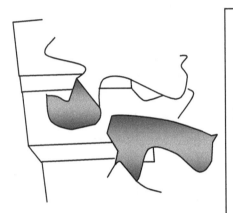

Oblique image of vertebra
Darkened part represents the "Scotty dog," complete with a fracture <u>and</u> separation at the site of the pars interarticularis. This is a beheaded Scotty dog, and is the classic finding for a Type II spondylolisthesis, subtypes A and C.

Type III   Degenerative Spondylolisthesis   Degenerative changes to the zygopophyseal joints, secondary to chronic segmental instability. Is much more common in females than males, and is more common in blacks than whites. The most common site is at L4. It is not seen in those less than 40 years old, is rare between ages 40 and 50, and becomes more prevalent in those above 50 years of age.

Type IV   Traumatic Spondylolisthesis   This is characterized by fractures anywhere in the intervertebral bony connections except the pars interarticularis; it invariably heals with immobilization. It is the least common type of spondylolisthesis.

Type V   Pathologic Spondylolisthesis   This is due to bone disease secondary to such disorders as cancer, Paget's disease, osteogenesis imperfecta, etc. Even though it is the second least common type of spondylolisthesis, it is still a rare cause for spondylolisthesis.

Ankylosis
  ❖ An abnormal immobility of the joint; if it is a true ankylosis (a bony ankylosis),
    the immobility is due to the fusion of the bones that form the joint.
  ❖ Causes restriction of movement; can also be accompanied by muscle spasms and
    achiness if neural structures are impinged.

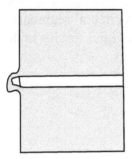

Ankylosis of Two
Vertebrae

Spondylosis
  ❖ Degenerative changes of the intervertebral discs accompanied by ankylosis of
    adjacent vertebral bodies, leading to a narrowed spinal canal or neural foramen.
  ❖ Causes chronic achy low back pain

Osteoarthrosis
  ❖ A synonym for degenerative joint disease (DJD) or osteoarthritis, which are
    themselves synonyms for one another
  ❖ Causes chronic achy focal pain that is exacerbated by rest and relieved by
    non-weight-bearing or non-jarring movement

Spinal stenosis
  ❖ Spinal canal narrowing, causing pressure on the nerve roots or the cord
  ❖ The most common form is due to osteoarthrosis in which degenerative arthritic
    changes occur in the facet joint with concomitant changes in the intervertebral disc
  ❖ Causes achy low back pain with sharp, shooting pain that often radiates into the
    lower extremity; the pain is exacerbated by standing and walking and backward
    bending

Flexion contracture of the Iliopsoas
  ❖ Usually precipitated and/or exacerbated by long-held positions with the psoas
    shortened (the leg flexed), such as in the position held while weeding.
  ❖ An extremely common somatic disorder

The **most common congenital anomaly in the lumbar** region is asymmetry of the joint facets, known as <u>**zygopophyseal tropism.**</u>

The **second most common congenital anomaly of the lumbar** area is <u>**sacralization.**</u>

<u>**Important Anatomical Landmarks**</u>
Iliac crest is at the horizontal level of L4-L5 vertebral segments.
The umbilicus is at the level of L3-4 segments (and resides in the T10 dermatome).

Chapter 4 Review Cases

1. A 34-year old obese male presents with complaints of low back pain. Suspecting a herniated disc, you order imaging wherein a herniated L3 intervertebral disc is identified. What nerve root risks impingement?
    a. L1 nerve root
    b. L2 nerve root
    c. L3 nerve root
    d. L4 nerve root
    e. L5 nerve root

2. The previous patient refuses further evaluation and only wants pain control. He goes home and returns one month later in excruciating pain. He describes it as being the same type of pain and in the same location as a month previous, but having been exacerbated this last time by picking up a cinder block. If this herniation has gotten larger, and the amount of herniated material is greater, what nerve roots are in most jeopardy of impingement?
    a. L3 nerve root alone
    b. L2 and L3 nerve roots
    c. L4 nerve root alone
    d. L4 and L5 nerve roots
    e. L5 nerve root alone

3. A 58 year-old construction worker presents with neck pain and is later found to have a herniation of the C5-C6 intervertebral disc. The herniation is massive. What nerve roots do you most expect to be impinged?
    a. C4 nerve root alone
    b. C4 and C5 nerve roots
    c. C5 nerve root alone
    d. C5 and C6 nerve roots
    e. C6 nerve root alone

4. A 45 year old female presents with severe lumbar lordosis. What might be an expected Ferguson's angle?
    a. 5 degrees
    b. 10 degrees
    c. 15 degrees
    d. 30 degrees
    e. 40 degrees

5. An infant is examined for evident anomalies of the back. In the area under investigation, imaging reveals a sac of cerebrospinal fluid emanating from the posterior area of L4. No nerve roots are in the sac. What is the diagnosis?
   a. Spina bifida occulta
   b. Spina bifida meningocele
   c. Spina bifida meningomyelocele
   d. Spina bifida cystica
   e. Rachischisis

6. A patient has L5FRRSR. What sacral somatic dysfunction may this cause?
   a. Sacrum rotated right on a right sacral oblique axis
   b. Sacrum rotated left on a left sacral oblique axis
   c. Sacrum rotated right on a left sacral oblique axis
   d. Sacrum rotated left on a right sacral oblique axis
   e. Sacrum flexed bilaterally

7. A 62-year old patient was in a head-on motor vehicle accident while sitting in the passenger seat and reaching to pick something up off the floor. He complains of severe, sharp pain in his low back, along with the inability to feel his groin. What is the likely diagnosis?
   a. Herniated nucleus pulposus
   b. Cauda equina syndrome
   c. Spondylolysis
   d. Spondylolisthesis
   e. Ankylosis

8. A 32-year old female is suffering from sharp, stabbing low back pain. It started abruptly after having bent over to pick up her 60 pound child. What is the most likely diagnosis?
   a. Herniated nucleus pulposus
   b. Cauda equina syndrome
   c. Spondylolysis
   d. Spondylolisthesis
   e. Osteoarthritis

9. A middle-aged female spent 2 hours hunched over wiping down her baseboards. When finished, she was unable to stand up straight. What is the likely diagnosis?
   a. Herniated nucleus pulposus
   b. Cauda equina syndrome
   c. Spondylolysis
   d. Spondylolisthesis
   e. Iliopsoas flexion contracture

10. A 43-year old male falls from a rock on which he was standing, injuring his low back. Initially, there was a brief sharp pain. That was two weeks ago. Now, he finally presents because he has had ongoing chronic, achy low back pain since the event, pain that is worsened anytime he tries to walk or be active. It feels better when he rests. Lateral lumbar X-rays show excellent alignment of the vertebrae in the area of concern. What is the most probably diagnosis?
    a. Herniated nucleus pulposus
    b. Cauda equina syndrome
    c. Spondylolysis
    d. Spondylolisthesis
    e. Spondylosis

11. What imaging position is best to use to confirm the diagnosis for the previous patient?
    a. Anterior-posterior
    b. Posterior-anterior
    c. Lateral
    d. Oblique
    e. Superior

12. A patient is determined to have L3 forward slipped about 50% of the way. What is the diagnosis?
    a. Grade 1 spondylolysis
    b. Grade 1 spondylolisthesis
    c. Grade 2 spondylosis
    d. Grade 2 spondylolisthesis
    e. Grade 3 spondylolisthesis

13. A 7-year old girl is brought by her parents for evaluation. She has been complaining for many months about chronic achy low back pain. She started ballet lessons this year, and finds it difficult to go to class because the lessons aggravate her back. No beheaded or collared Scotty dog is seen on imaging, but a lateral lumbar X-ray reveals a significant anomaly. What is the most probable diagnosis?
    a. Spinal stenosis
    b. Dysplastic spondylolisthesis
    c. Traumatic spondylolisthesis
    d. Pathologic spondylolisthesis
    e. Spondylolysis

14. A patient demonstrates a beheaded Scotty dog on oblique lumbar X-ray. What is the diagnosis?
    a. Spondylolysis
    b. Spondylosis
    c. Type II, subtype A spondylolisthesis
    d. Isthmic spondylolisthesis, subtype B
    e. Degenerative spondylolisthesis

15. A 45-year old overweight, deconditioned female wants to be evaluated for back pain she claims to have had for 5 years. She notes that it is getting worse, and is particularly exacerbated by rest. Whenever she sits or lies down for a long period of time, it begins to ache unbearably. Getting up in the morning is the worst time. However, just in the past couple of months, she also has had some severe pain that occurs whenever she carries anything. All of the pain is located to her low back. What is the likely diagnosis?
    a. Osteoarthritis
    b. Herniated nucleus pulposus
    c. Spondylosis
    d. Spondylolisthesis
    e. Spondylolysis

16. Where will you place your hands to allow your thumbs to meet bilaterally, allowing your thumbs to identify the location you will insert the needle for a lumbar puncture?
    a. PSIS's
    b. ASIS's
    c. Sacral base
    d. Iliac crests
    e. Ischial tuberosities

Answers to Chapter 4 Cases

1. D   Although any lumbar nerve root leaves the spinal column at the level of the same numbered vertebral segment, it does so at a location above the same-numbered intervertebral disc. Thus, a herniation of the L3 disc, considering that lumbar herniations go posteriorly laterally (and not superiorly), will not be able to generally impinge the L3 nerve root; the nerve root comes out above the disc. Hence, the next nerve root in line for exit ends up being impinged, and that is the L4 nerve root in this case.

2. D   Unlike other areas of the spine, the larger the disc herniation in the lumbar spine, the more nerve roots will be impinged. This is because, in the lumbar spine, disc herniations are directed posteriorly laterally, directly into the area of the nerve roots and the cauda equina. The larger the herniated mass, the more nerve roots are impinged. Thus, L4 will still be impinged, by L5 will likely also be impinged, too. If the herniation worsens, S1, S2, and so forth can also become impinged by this disc.

3. E   In the cervical spine, there is not a cauda equina, so more than one nerve root cannot be impinged by a single herniated nucleus pulposus, no matter how large it is. The C5-6 intervertebral disc is the same as the C5 disc, the intervertebral disc that is positioned between C5 vertebral segment and C6 vertebral segment. Nerve roots in the cervical region leave the spine at a position higher than the same numbered vertebral segment. Thus, a C6 nerve root comes out above C6 vertebral segment. The C5 disc is in that location. Thus, herniation of the C5 disc may result in impingement of the C6 nerve root. If it is a large herniation, it will not impinge any more nerve roots other than C6 – a situation quite different than that seen in the lumbar spine.

4. E   Lordosis is a lumbar curvature that is an extreme version of the normal lumbar curvature; hence, lordosis causes the buttocks region to protrude out, the abdomen to be position anterior to the center of gravity, and the thoracic spine to be "swayed" back (more posterior) than usual (which is why lordosis is also known as swayback). In any case, this is accompanied by a sacrum that flexes more than usual, creating a larger lumbosacral angle (Ferguson's angle). A patient with severe lumbar lordosis will invariably, therefore, have a larger than normal Ferguson's angle. Normal for such an angle is 25-35 degrees; so it is conceivable that this patient would have a 40 degree Ferguson's angle.

5. B   This is spina bifida, a condition representing failure of closure of the laminae. As a result, material can emanate from the opening. In this case, there is a sac of CSF. The sac is surrounded by meninges. Thus, the only tissue that is outside of the vertebral canal is the meninges. Hence, this is a spina bifida meningocele. If nerve roots were in the sac, it would have been a spina bifida meningomyelocele. It should be noted that spina bifida cystica is a correct choice, too, since both a meningocele and a meningomyelocele represent two different versions of spina bifida cystica. However, the more accurate answer since it is specific to the diagnosis is spina bifida meningocele.

6. D   L5 can have profound effects on the sacrum. Accordingly, if L5 is rotated, it can cause the sacrum to rotate in the opposite direction. Likewise, if L5 is sidebent, it can induce an ipsilateral sacral oblique axis. Therefore, if L5 is rotated right, the sacrum can be induced to undertake left rotation; if L5 is sidebent right, the sacrum can assume a right sacral oblique axis. Thus, L5 FRRSR can cause the sacrum to rotate left on a right sacral oblique axis.

7. B   Flexion and downward stress on the spine encourages herniation of lumbar intervertebral discs. Certainly, the greater the flexion and the larger the downward stress, such as what would be experienced while flexing with a head-on motor vehicle collision, the greater the opportunity for massive central disc herniation. Such massive herniation is cauda equina syndrome, and results in the sharp, sudden low back pain associated with any lumbar disc herniation, along with saddle anesthesia (anesthesia of the groin and inner thigh) and fecal and/or urinary incontinence.

8. A   Flexion and/or downward pressures on the spine encourage lumbar disc herniation. This woman was both flexing and lifting a heavy weight. The classic symptoms of a disc herniation include sudden onset, sharp, intense pain in the area of the herniation (which is typically aggravated by flexing or stressing the area again). The history and the symptoms suggest herniated nucleus pulposus. It is unlikely to be cauda equina because, although the history and presenting symptoms are the same, there is also saddle anesthesia and fecal and/or urinary incontinence with the latter (symptoms patients rarely forget to mention if that is occurring).

9. E   Long-term deep hip flexion can induce a contracture of the iliopsoas, causing a temporary but painful inability to straighten the leg at the hip.

10. C   Chronic achy low back pain, exacerbated by activity, is classic for spondylolysis, spondylolisthesis, osteoarthritis, and spondylosis; ankylosis and spinal stenosis also demonstrate this pattern of pain, but the former is frequently accompanied by spasms and spinal stenosis is accompanied by concomitant sharp pains. Anyway, this patient's sole complaint at this time is chronic achy low back pain, limiting the primary differential diagnosis to spondylolysis, spondylolisthesis, osteoarthritis, and spondylosis. Osteoarthritis' achy pain is relieved by activity, not worsened – besides, it is a chronic condition that does not occur over several weeks. So, it is not osteoarthritis. Spondylosis, while marked by achy chronic low back pain, is also genuinely a chronic condition, taking years to develop. Thus, the only conditions under consideration are spondylolysis and spondylolisthesis. The lateral lumbar X-ray demonstrates normal vertebral alignment, thereby ruling our spondylolisthesis. Therefore, this patient has spondylolysis, caused by his fall from the rock.

11. D   An oblique lumbar X-ray will allow for evaluation of the pars by assessment of the Scotty dog. A collar on the Scotty dog will be demonstrated.

12. D   Forward slippage of a vertebral body is known as spondylolisthesis. 50% slippage falls into the categorization of a Grade 2 event.

13. B   Chronic back pain in a child is unusual and always warrants serious attention. Chronic achy low back pain is typical of spondylolysis, spondylolisthesis, spondylosis, and osteoarthritis (and, along with other symptoms, ankylosis and spinal stenosis). Spondylosis and osteoarthritis require years to develop, and so are not expected findings in a child. This child likely has either spondylolysis or spondylolisthesis. Since no anomalies of the pars are noted on imaging, spondylolysis is unlikely. Instead, a lateral X-ray reveals a significant anomaly – most likely a forward slippage. Thus, the likely diagnosis is spondylolisthesis. Young girls are most at risk for developing dysplastic spondylolisthesis, a spondylolisthesis that arises from a defect with which they are born. It tends to become symptomatic at ages 7-10. Those symptoms are typically described as chronic achy low back pain exacerbated by activity, symptoms consistent with any type of spondylolisthesis.

14. C   A beheaded Scotty dog indicates that there has been fracture at the pars interarticularis and subsequent separation of bony structures at that point. Such separation yields a spondylolisthesis. Any pathology of the pars causing a spondylolisthesis is known as an isthmic spondylolisthesis, otherwise known as a Type II spondylolisthesis. The type of isthmic spondylolisthesis characterized by separation at the fractured pars is subtype A or subtype C, both caused by different events but yielding the same result.

15. A   Being overweight puts tremendous strain on the weight-bearing joints of the body, particularly the lumbar spine, the hips, and the knees. Such strain can promote the development of osteoarthritis. Being deconditioned also promotes its development since lack of muscle tone increases the strain, along with the wear and tear, to joints. Thus, by history, this woman appears to have osteoarthritis. It takes years to develop and, once developed, tends to worsen with time. It is always characterized by achy chronic pain that is exacerbated by rest. The longer the person is immobile, the greater the ache becomes. This is known as gelling. This feature exists at all stages of the disease. However, as the joint damage progresses, eventually, in addition to the gelling, the patient also starts to suffer from sharp pain at the joint during use, particularly weight bearing use, of that joint. The patient complains of chronic low back pain that she had for years, that has worsened with time, that is exacerbated with rest, and that, most recently, also has evolved to additionally include sharp pain during activity. This is classic for osteoarthritis.

16. D   The site for insertion of the needle for a lumbar puncture (colloquially known as a spinal tap) should be at the L4-L5 interspace. That interspace is located on the same horizontal plane as the iliac crests. So, typically, we rest our hands on the iliac crests to allow our thumbs to extend medially bilaterally on the same horizontal plane, allowing us to locate the L4-5 interspace.

Chapter Five: SACRUM AND INNOMINATES

The sacrum is comprised of five fused vertebral elements. The most superior anterior portion of the sacrum is termed the sacral promontory, and corresponds with the anterior portion of S1. The most superior aspect of the sacrum is considered to be the base, while the inferior part is the apex.

The sacrum itself contains four foramina bilaterally that permit passage of the sacral nerves. In the inferior posterior aspect of the sacrum can be found the sacral hiatus, an opening which represents failure of S5 laminal closure. Here is the site of performance of sacral epidural nerve blocks (caudal analgesia). The sacral cornua, which are the inferior articular processes of S5, project inferiorly bilaterally on each side of the sacral hiatus. These cornua serve as a helpful guide in finding the sacral hiatus.

There are 2 innominates in any individual, and each innominate is comprised of three bones: the ilium, the ischium, and the pubic bone, all of which are joined at the acetabular notch. These bones are partially cartilaginous at birth and do not fully ossify and fuse until around age 20.

The Sacrum and Innominate

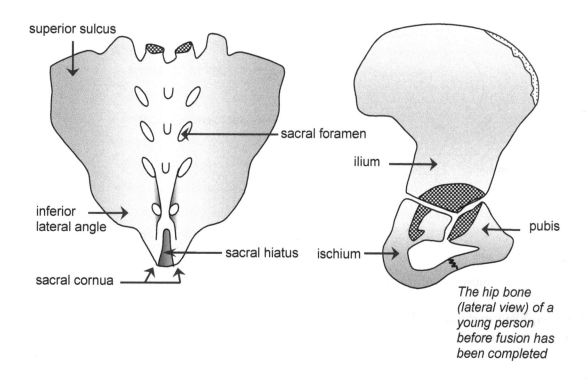

superior sulcus

sacral foramen

ilium

inferior lateral angle

sacral hiatus

ischium

pubis

sacral cornua

The hip bone (lateral view) of a young person before fusion has been completed

Articulations and Ligaments

The innominates have several joints: one at the pubic symphysis, two with the femurs, and two sacroiliac (SI) joints. The innominates articulate with the sacrum at the **sacroiliac (SI) joints** and with the femurs at the **acetabula.** The two pubic bones articulate with one another at the **pubic symphysis.** Besides articulating with the innominates, the sacrum also articulates with the coccyx at the **sacrococcygeal joint** as well as with the lumbar spine at L5. The stability of the pelvic and hip region is augmented by various ligaments. The main ligaments are the **sacrotuberous ligaments, sacrospinous ligaments, iliolumbar ligaments, and sacroiliac ligaments.** The former three are considered accessory ligaments while the sacroiliac ligaments are considered true ligaments. The sacrotuberous ligament ligates the sacrum to the ischial tuberosity while the sacrospinous ligament connects the sacrum to the ischial spines. The sacrospinous ligament is anatomically significant in that it divides the **greater and lesser sciatic foramen.** The iliolumbar ligaments attach to L4, L5, the iliac crests, and the sacroiliac (SI) joints. They are usually the first ligaments to become tender when lumbosacral dysfunction occurs. There are three sacroiliac ligaments: the anterior, posterior, and interosseous SI ligaments, all of which aid in the stabilization of the SI joints.

A Note About the Coccyx: just anterior to the coccyx lies the ganglion impar, the site at which the right and left sympathetic chains join.

Ligaments of the Lumbosacral Region and Innominates

This is a depiction of the left side of the sacrum and the left innominate from a posterior view.

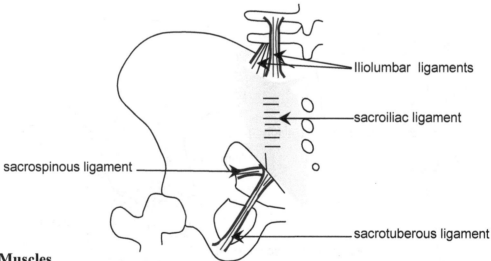

Iliolumbar ligaments

sacroiliac ligament

sacrospinous ligament

sacrotuberous ligament

Muscles

There are several muscles of importance in the area of the sacrum and innominates.

Pelvic diaphragm: comprised of the levator ani (the puborectalis, pubococcygeus, and iliococcygeus muscles) and the coccygeus muscles. It functions to support pelvic viscera and the pelvic floor, it aids in raising the pelvic floor, and it serves to constrict the lower rectum and vagina.

Piriformis: Originates at the inferior anterior aspect of the sacrum and inserts onto the greater trochanter of the femur. It functions to 1) externally rotate the thigh, 2) extend the thigh, and 3) abduct the thigh when the hip is flexed. It is innervated by S1 and S2.

Clinical Note: A little over 10% of the population has the peroneal portion of the sciatic nerve running *through* the belly of the piriformis. This may result in buttock pain secondary to piriformis hypertonicity as well as sciatic nerve-related pain radiating down the leg. The latter pain is described as sciatica.

Iliopsoas: The psoas attaches to the sides of T12 to L5 vertebrae and joins the iliacus within the true pelvis. The iliacus attaches to the iliac crest, iliac fossa, sacral ala, and anterior sacroiliac ligaments. They pass together toward the leg, whereupon the psoas attaches to the lesser trochanter of the femur and the iliacus attaches to the psoas major tendon and to the body of the femur inferior to the lesser trochanter. The psoas is innervated by branches of the lumbar plexus, namely L1-3. The iliacus is innervated by the femoral nerve (L2 and L3). The iliopsoas acts to flex the thigh at the hip joint and to stabilize that joint.

> TRUE PELVIS: a funnel-shaped area bounded anteriorly by the pubic bones and posteriorly by the sacrum.
> FALSE PELVIS: technically is part of the abdomen, and is located in the area medial to the ilia.

Clinical Note: A flexion contracture of the iliopsoas may cause lumbar vertebrae, most commonly L1 and/or L2, to be flexed, as well as sidebent and rotated to the side of the contracture.

Nerves

Lumbar plexus: formed from contribution from T12 to L4, and innervates the muscles of the abdomen and thigh, including the iliopsoas, quadriceps, adductor group, sartorius, and gracilis. The lumbar plexus also provides sensation to the thigh, buttocks, lower abdomen, and pubic area.

Sacral plexus: formed from L4-5, S1-3, and a portion of S4. The sacral plexus provides motor and sensory innervation to the pelvis and lower extremity as well as parasympathetic innervation (S2,3,4) to the left colon and pelvic organs. Nerve branches of the sacral plexus include the sciatic, pudendal, and superior and inferior gluteal nerves.

Sciatic nerve: a branch of the sacral plexus, and originates from L4-S3. The sciatic nerve usually runs through the greater sciatic notch just below the piriformis muscle; other anomalies exist, including sciatic nerves that run through the muscle itself.

Ganglion impar: sits anterior to the coccyx, and is the fusion of the left and right sacral sympathetic trunks.

Sacral Mechanics

There are four types of sacral motion:

Inherent Motion: the inherent motion of the sacrum associated with the cranial cycle of flexion and extension. During craniosacral extension, the sacrum "dips forward" into flexion, and is said to nutate. Conversely, during craniosacral flexion, the sacrum "tips backward" into extension, and is said to counternutate. Nutation and counternutation are terms used to describe sacral movement with special respect to craniosacral mechanics. The axis of motion for the sacrum is a horizontal axis that runs through S2.

Respiratory Motion: the motion the sacrum undertakes during inhalation and exhalation. This motion occurs around a horizontal axis that runs through S2, the same axis as that for inherent motion. With inhalation, the sacral base moves posteriorly while, during exhalation, the base moves anteriorly.

Postural Motion: the motion undertaken by the sacrum during flexion or extension of the torso. With flexion of the torso, the sacral base moves posteriorly. Conversely, the sacral base moves anteriorly with torso extension. The axis for this motion is a transverse axis through S3.

Dynamic Motion: the sacral motion that occurs with ambulation. When one steps forward with the left leg, thereby placing the body's weight onto the right leg, a right sacral oblique axis will be engaged, and vice versa.

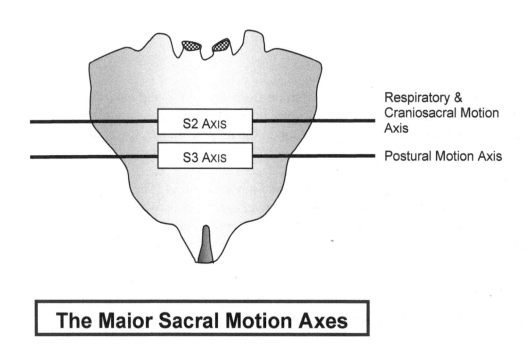

The Major Sacral Motion Axes

The Sacral Oblique Axes

The posterior view of the sacrum is depicted

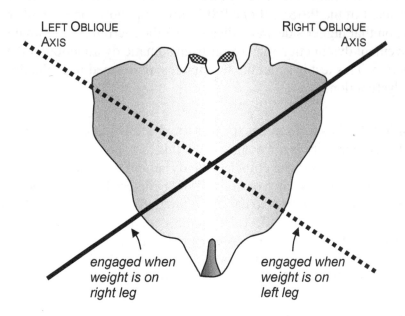

LEFT OBLIQUE AXIS

RIGHT OBLIQUE AXIS

engaged when weight is on right leg

engaged when weight is on left leg

THE 2 L5-SACRAL RULES
How L5 dysfunction affects sacral positioning:

1.   L5 rotation is opposite of sacral rotation. For example, if L5 is rotated right, the sacrum will be rotated left.
2.   L5 sidebending engages an ipsilateral sacral oblique axis. For example, if L5 is sidebent left, a left sacral oblique axis is engaged.

When referring to rotation and oblique axes of the sacrum, convention states that one refer to the rotation being "on" the axis. For example, a sacrum rotated left with a right oblique axis is actually rotated left ON a right oblique axis; in other words, it is a left on right, denoted as "L on R".

Tests for Sacral and Innominate Somatic Dysfunction

Standing Flexion Test: This test screens for somatic dysfunction of the **innominates, sacrum, and lower extremity.** To perform it, the clinician stands behind the standing patient (whose feet are only about 4 - 6 inches apart) and places his thumbs slightly inferior to the PSIS's bilaterally. Then the patient is instructed to bend forward, allowing his/her arms to dangle towards the floor. While the patient it bending forward, the clinician takes note of any asymmetry in movement of the PSIS's. If one PSIS moves up more than the other, the test is considered positive on that higher side and indicates that the sacrum is not moving properly on that side. Such movement dysfunction may result from somatic dysfunction of the leg, innominates, or sacrum. This test is generally the first test performed in evaluating possible innominate/sacral dysfunction.

Seated Flexion Test: This test screens for somatic dysfunction of the **sacrum and/or the innominates;** the lower extremity is not evaluated by this test. To perform this, the clinician instructs the patient to sit down and the clinician stands behind the seated patient. Then, the clinician places his thumbs slightly inferior to the PSIS's bilaterally and instructs the patient to bend forward, allowing the his/her arms to dangle to the floor. The clinician takes note of any asymmetry in PSIS movement. If the PSIS rides up more on one side than the other, the test is considered positive on that higher side and indicates sacroiliac dysfunction. This test is generally performed after the standing flexion test. Be aware that some authorities consider a positive seated flexion test as indicative of only sacral and not innominate dysfunction!!!

Lumbosacral Spring Test: Also known as the **Lumbar Spring Test,** this evaluates for increased or decreased lumbar lordosis as well as for sacral base flexion/extension. The patient assumes the prone position. The clinician places the heel of one hand over the lumbosacral junction, and the heel of the other hand over that first hand. A gentle, quick downward pressure is exerted at the lumbosacral junction with the clinician's hands, and is done two or three times in rapid succession. The clinician notices the <u>quality</u> of springing motion. If there is good springing, the test is negative. Poor or no springing renders the test positive, and indicates that the lumbosacral junction is restricted. Such restriction occurs in any dysfunction in which the sacral base is unilaterally or bilaterally positioned posteriorly.

Innominate Rocking Test: This is also known as the **ASIS Compression Test,** and is often used as a confirmatory test in assessing flexion test findings. It helps to identify the side of dysfunction of the <u>sacrum, innominate,</u> or <u>pubic symphysis.</u> The patient assumes the supine position. The clinician places the palmar aspect of his hands over the anterior superior iliac spines (ASIS's) bilaterally. Then, downward compression is applied unilaterally while the other side is stabilized. This is performed on each side to assess sacroiliac (SI) joint stability and laxity. If there is resistance to rocking, the test is positive for that side and indicates pelvic somatic dysfunction that side.

INNOMINATE SOMATIC DYSFUNCTION

Perform a standing flexion test. The side of the positive test will be the side of the dysfunction, unless there is no iliosacral motion (whereupon a seated flexion test must be performed to rule in various sacral dysfunctions or a false positive standing flexion).

dysfunction	findings	etiology
Anterior innominate rotation (one innom. rotates anteriorly with respect to the other)	ASIS inferior ipsilaterally PSIS superior ipsilaterally longer leg ipsilaterally positive innominate rocking test ipsilaterally positive standing flexion test ipsilaterally (+/- seated flexion test ipsilaterally)	tight quadriceps rectus femoris or adductor group dysfunction

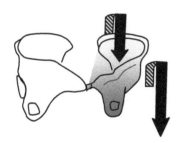

Symptoms: hamstring tightness, sciatica (due to piriformis dysfunction)

dysfunction	findings	etiology
Posterior innominate rotation (one innom. rotates posteriorly with respect to the other)	ASIS superior ipsilaterally PSIS inferior ipsilaterally shorter leg ipsilaterally positive standing flexion test ipsilaterally positive innominate rocking test ipsilaterally (+/- seated flexion test ipsilaterally)	tight hamstrings

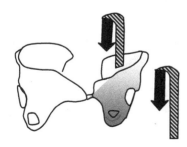

Symptoms: groin pain and/or medial knee pain (due to sartorius dysfunction)

| dysfunction | findings | etiology |

Superior innominate shear
"superior innominate subluxation" or "upslipped innominate" (one innom. is positioned more superiorly than the other)

ASIS superior ipsilaterally
PSIS superior ipsilaterally
pubic ramus superior ipsilaterally
positive innominate rocking test
    ipsilaterally
positive standing flexion test
    ipsilaterally
(+/- seated flexion test ipsilaterally)

a fall on or thrust up the ipsilateral gluteal area, surprise step off curb into a hole

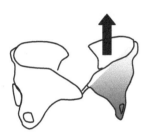

Symptoms: pelvic pain or contralateral SI joint pain and/or low back pain

Inferior innominate shear
"inferior innominate subluxation" or "down-slipped innominate" (one innom. is positioned more inferiorly than the other)

ASIS inferior ipsilaterally
PSIS inferior ipsilaterally
pubic ramus inferior ipsilaterally
positive innominate rocking test
    ipsilaterally
positive standing flexion test ipsilaterally
(+/- seated flexion test ipsilaterally)

trauma/car accident

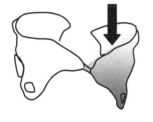

Symptoms: pelvic pain or contralateral SI joint pain

dysfunction	findings	etiology

Superior pubic shear
(one pubic bone is superior to the other)

pubic bone is superior
    ipsilaterally
PSIS is level or inferior
    ipsilaterally
ASIS is level or superior
    ipsilaterally
positive standing flexion test
    ipsilaterally
positive ASIS compression test
    ipsilaterally
(+/- seated flexion test ipsilaterally)

tight rectus abdominus muscle, unusual trauma, innominate rotation (posterior), third trimester pregnancy or delivery

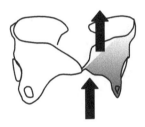

Symptoms: tender inguinal area with or without dysuria or frequency

Inferior pubic shear
(one pubic bone is inferior to the other)

pubic bone is inferior
    ipsilaterally
PSIS is level or superior
    ipsilaterally
ASIS is level or inferior
    ipsilaterally
positive standing flexion test
    ipsilaterally
positive ASIS compression tcst
    ipsilaterally
(+/- seated flexion test ipsilaterally)

tight adductor muscles, unusual trauma, innominate rotation (anterior), third trimester pregnancy or delivery

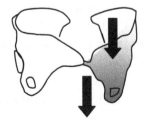

Symptoms: pelvic or SI joint pain

dysfunction	findings	etiology
Innominate inflare (innominate is rotated medially)	ASIS more medial ipsilaterally taut posterior pelvic muscles ipsilaterally positive standing flexion test ipsilaterally positive ASIS compression test ipsilaterally (+/- seated flexion test ipsilaterally)	trauma

Symptoms: sometimes dysuria and/or frequency

dysfunction	findings	etiology
Innominate outflare (innominate is rotated laterally)	ASIS more lateral ipsilaterally lax posterior pelvic muscles ipsilaterally positive standing flexion test ipsilaterally positive ASIS compression test ipsilaterally (+/- seated flexion test ipsilaterally)	trauma

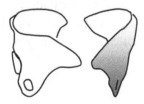

Symptoms: pelvic or SI joint pain

SACRAL SOMATIC DYSFUNCTION

Two important, but antiquated, terms one should be familiar with are "anterior sacrum" and "posterior sacrum". An **anterior sacrum** is a sacrum that is flexed and sidebent to the opposite side of rotation. A **posterior sacrum** is a sacrum that is extended and sidebent to the opposite side of rotation. Also, there is a sacral dysfunction called a **sacral margin posterior,** whereby one side (or "one margin") of the sacrum from base to apex is positioned more posteriorly than its other side. This dysfunction is not recognized by all in the field, and so is only discussed herein to be complete. The following, however, is the more modern and now conventional method of describing the various sacral dysfunctions:

Handbook of OMT Review

There are <u>Three Types of Sacral Dysfunction:</u>

1) Bilateral sacral flexion/extension = the sacral base is either anterior (flexed) or posterior (extended)
2) Sacral shears = the sacral base is more anterior or posterior unilaterally
3) Sacral torsions = rotation of the sacrum about an oblique axis in combination with L5 somatic dysfunction.

<u>dysfunction</u>	<u>findings</u>	<u>etiology</u>

Bilateral sacral flexion
"Sacral Base Anterior"
(the sacral base is positioned more anteriorly)

deep R and L superior sulci
shallow R and L ILAs
increased lumbar lordosis
lumbosacral spring test negative
restricted ILA springing bilaterally
false negative seated and standing
       flexion tests

delivery or
trauma

Symptoms: low back pain that worsens with backward bending

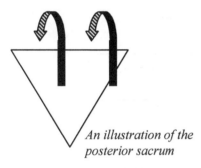
An illustration of the posterior sacrum

> NOTE: This dysfunction is <u>very</u> common in postpartum females

Bilateral sacral extension
"Sacral Base Posterior"
(the sacral base is positioned more posteriorly)

shallow R and L superior
      sulci
deep R and L ILAs
decreased lumber curvature
lumbosacral spring test positive
restricted superior sulci springing
      bilaterally
false negative seated and standing
      flexion tests

trauma

Symptoms: low back pain that worsens with forward bending

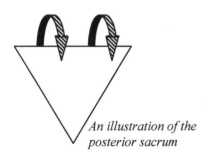
An illustration of the posterior sacrum

SACRAL SHEARS: unilateral sacral flexion/extension

<u>dysfunction</u>   <u>findings</u>   <u>etiology</u>

Unilateral sacral flexion
"Unilateral Anterior Sacral Shear"
(sacral base is anterior
unilaterally)

deep superior sulcus
    ipsilaterally
shallow ILA ipsilaterally
restricted ILA springing
    ipsilaterally
negative lumbosacral spring
    test
positive seated flexion test
    ipsilaterally
positive standing flexion test
    ipsilaterally
shorter leg ipsilaterally

surprise step
into hole or
off curb, or
lumbar
dysfunction

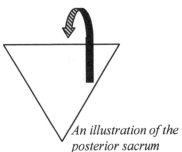
*An illustration of the
posterior sacrum*

Symptoms: low back pain

NOTE: This is the most common
type of sacral shear!

Unilateral sacral extension
"Unilateral Posterior Sacral Shear"
(sacral base is posterior deep
unilaterally)

shallow superior sulcus
    ipsilaterally
deep ILA ipsilaterally
positive lumbosacral spring
    test
positive seated flexion test
    ipsilaterally
positive standing flexion test
    ipsilaterally
longer leg ipsilaterally

trauma or
lumbar
dysfunction

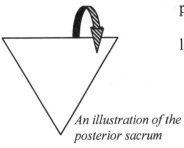
*An illustration of the
posterior sacrum*

Symptoms: low back pain

SACRAL TORSIONS: Forward sacral torsion, Backward sacral torsion

- implies that the sacrum rotates in one direction while the lumbar spine rotates in the opposite direction; the resulting rotation of the sacrum occurs about an oblique (diagonal) axis. This rotation around a sacral oblique axis is also considered a torsion.

dysfunction

findings

etiology

Forward sacral torsion
(rotation is in the same direction as the axis)

There are two types:
L on L *or*
R on R

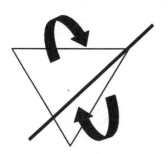

An illustration of the posterior sacrum

deep superior sulcus
    contralateral to
    rotational
    component
shallow ILA ipsilateral to
    rotational component
lumbar sidebent ipsilateral
    to oblique axis
restricted ILA springing
    ipsilateral to rotational
    component
lumbar rotation in opposite
    direction of sacral
    rotation
positive seated flexion test
    contralateral to
    oblique axis
positive standing flexion test
    ipsilateral to positive
    seated flexion test
negative lumbosacral spring test

trauma or
lumbar
dysfunction

Symptoms: low back pain with SI, inguinal or groin pain; the low back pain is exacerbated by backward bending

dysfunction	findings	etiology

Backward sacral torsion
(rotation is in the opposite side of
the axis, i.e. left rotation on a right
oblique axis)

There are two types:
R on L *or*
L on R

*An illustration of the
posterior sacrum*

shallow superior sulcus
    ipsilateral to
    rotational
    component
deep ILA contralateral to rotational
    component
lumbar sidebending to side ipsilateral of
    oblique axis
L5 rotated to the side
    contralateral to the
    rotational component
positive lumbosacral spring test
restricted superior sulcus springing
    ipsilateral to the rotational component
positive seated and standing flexion test
    on side contralateral to oblique
    axis

trauma or
lumbar
dysfunction

*Symptoms: low back pain or SI discomfort
that worsens with bending forward or
walking*

Clinical Note: Before treating a sacral dysfunction, any dysfunction in L5 should be identified
and treated first. L5 treatment usually spontaneously resolves any sacral dysfunction.

Chapter 5 Review Cases

1. A 39-year old man "twisted" his low back while playing tennis. What was the likely source of the initial pain?
   a. Nerve root compression
   b. Stretch of sacrospinous ligaments
   c. Stretch of sacroiliac ligaments
   d. Stretch of iliolumbar ligaments
   e. Sacral torsion

2. A 50-year old female develops cystocele. Laxity of what muscles has contributed to this?
   a. Piriformis
   b. Pelvic diaphragm
   c. Iliopsoas
   d. Gluteus maximus
   e. Rectus abdominus

3. You diagnose a patient with a right iliopsoas flexion contracture, with a resulting lumbar somatic dysfunction at L1, L2, and L3. What is the specific dysfunction noted with those vertebral segments?
   a. L1-L3 FRRSR
   b. L1-L3 NSRRL
   c. L1-L3 FRLSL
   d. L1-L2 NSLRR
   e. L1-L2 FRRSR

4. L4 nerve root pathology can result in pathology of which of the following structures?
   a. Lumbar plexus
   b. Lumbar plexus and sacral plexus
   c. Sacral plexus
   d. Sacral plexus and sciatic nerve
   e. Lumbar plexus, sacral plexus, and sciatic nerve

5. A patient has a left psoas syndrome. What pelvic shift and sacral dysfunction is expected, as a result?
   a. Right pelvic shift with a right on left sacral torsion
   b. Right pelvic shift with a left on left sacral torsion
   c. Right pelvic shift with a right on right sacral torsion
   d. Left pelvic shift with a left on right sacral torsion
   e. Left pelvic shift with a left on left sacral torsion

6. With respect to the previous patient, besides pain directly caused by the iliopsoas contracture, what other complaints is this patient likely to have?
   a. Left sciatica with left gluteal and posterior thigh pain
   b. Right sciatica with left gluteal and posterior thigh pain
   c. Left sciatica with right gluteal and posterior thigh pain
   d. Right sciatica with right gluteal and posterior thigh pain
   e. Right sciatica with right gluteal pain, and left posterior thigh pain

7. Which of the following individually will cause the sacral base to move anteriorly?
   a. Craniosacral extension, inhalation, torso flexion
   b. Craniosacral flexion, exhalation, torso extension
   c. Craniosacral extension, exhalation, torso extension
   d. Craniosacral flexion, inhalation, torso flexion
   e. Craniosacral extension, exhalation, torso flexion

8. A patient does not have sacral somatic dysfunction, but, physiologically, has a right sacral oblique axis. On what leg(s) is he standing?
   a. Both
   b. Right
   c. Left
   d. Neither
   e. It cannot be determined

9. A 48-year old male is evaluated for low back pain. The right transverse process of L5 is deeper than the left, and this asymmetry is most noticeable in neutral. What sacral somatic dysfunction most likely will result?
   a. None
   b. Right on right
   c. Right on left
   d. Left on right
   e. Left on left

10. A 27-year old male with a history of recently tripping on a hiking trip presents with complaints of low back pain accompanied with pain over the left sacroiliac joint. Physical exam reveals a superior left ASIS, PSIS, and pubic ramus. The patient complains of increased pain in the pelvic region during ambulation. There is a positive right ASIS compression test. What is the diagnosis?
    a. Right superior innominate shear
    b. Left inferior innominate subluxation
    c. Right upslipped innominate
    d. Left superior innominate shear
    e. Right inferior innominate subluxation

11. A 19-year old male athlete presents with sciatica. Physical exam reveals an inferior right ASIS, a superior right PSIS, and a positive right standing flexion test. What may have caused this dysfunction?
    a. Tight quadriceps
    b. Fall on the ipsilateral gluteal area
    c. Tight hamstrings
    d. Third trimester pregnancy
    e. Tight rectus abdominus

12. During an evaluation, a patient is found to have a lateral right ASIS following a traumatic event. A contralateral positive standing flexion test is demonstrated. What is the diagnosis?
    a. Left innominate inferior shear
    b. Right innominate inflare
    c. Left innominate inflare
    d. Right innominate outflare
    e. Left innominate outflare

13. Physical exam of a patient reveals a superior right pubic ramus and ASIS. The innominate rocking test is positive on the right. What other findings do you expect?
    a. Longer right leg
    b. Superior left ASIS
    c. Positive left standing flexion test
    d. Inferior right PSIS
    e. Tight right adductor muscles

14. On examination of a patient's sacrum, the following findings are demonstrated: deep right and left superior sulcus, shallow right and left ILAs, and negative seated and standing flexion tests. What are the expected lumbosacral spring test and lumbar spinal findings?
    a. Positive lumbosacral spring test, increased lumbar lordosis
    b. Positive lumbosacral spring test, decreased lumbar lordosis
    c. Positive lumbosacral spring test, no change in anterior-posterior lumbar curvature
    d. Negative lumbosacral spring test, no change in anterior-posterior lumbar curvature
    e. Negative lumbosacral spring test, increased lumbar lordosis

15. You palpate a patient's sacrum and find the left superior sulcus to be deep and the right inferior lateral angle to be shallow. Lumbosacral spring test is negative. There is a positive left seated and standing flexion test. What is the dysfunction?
    a. R on R
    b. R on L
    c. L on R
    d. L on L
    e. Bilateral sacral flexion

16. In evaluating a patient's sacrum, the right superior sulcus appears to be deeper than the left, and the left ILA is palpated to be more shallow than the right. The patient complains of low back pain, particularly when forward bending. There is a positive ASIS compression test on the left. What dysfunction exists?
    a. Left rotation on a left sacral oblique axis
    b. Left rotation on a right sacral oblique axis
    c. Right rotation on a left sacral oblique axis
    d. Right rotation on a right sacral oblique axis
    e. Sacral flexion on a horizontal axis at S2

Answers to Chapter 5 Cases

1. D   The iliolumbar ligaments are attached to L4, L5, the iliac crests, and the sacroiliac joints. They are usually one of the first generators of pain in cases of lumbosacral dysfunction, including minor cases of overstretching and temporary twisting.

2. B   The pelvic diaphragm, comprised of the levator ani and the coccygeus muscles, is responsible for adequate suspension of the rectum in males, as well as the urinary bladder, vagina, uterus, and rectum of the female. Laxity of the pelvic diaphragm can result in prolapse of these structures, yielding cystocele (prolapse of urinary bladder), rectocele (prolapse of rectum), and hystocele (prolapse of the uterus).

3. A   Iliopsoas flexion contractures "pull on" all of the vertebrae to which the psoas muscle attaches, specifically T12-L5. The two segments most likely to assume a dysfunction are L1 and L2, although any segment to which the psoas is attached can undertake a somatic dysfunction. Thus, L1-L3 can easily become dysfunctional. Iliopsoas contracture causes dysfunction to affected segments such that they become flexed, as well as sidebent and rotated to the same side of the iliopsoas contracture.

4. E   L4 nerve root contributes to the lumbar plexus, sacral plexus, and the sciatic nerve. While all answers are technically correct, the best answer is E because it is the most inclusive and accurate.

5. A   Psoas syndrome is another term for iliopsoas flexion contracture. The results of such a contracture are ipsilateral rotation and sidebending of any of the vertebrae to which the psoas muscle is attached (T12-L5), as well as flexion of the affected vertebrae. In addition, the contracture causes the pelvis (both innominates with the sacrum) to shift in position, specifically to shift to the side contralateral to the contracture. Hence, a left psoas syndrome will result in a right pelvic shift. The right pelvic shift with the Type II Fryette flexion of the lumbar vertebrae induces a sacral torsion, complete with an axis ipsilateral to the iliopsoas contracture (and ipsilateral to the lumbar sidebending). Because the lumbar vertebrae are pitched forward, the sacrum will dip backward around that axis, producing a backward sacral torsion. In this case, it will produce a backward sacral torsion with a left axis; since backward sacral torsions are always either right on left or left on right, we know that this dysfunction must be a right rotation on the left axis (R on L).

6. D   The pelvic and sacral dysfunction occur in conjunction with a resulting spasm of the piriformis on the side contralateral to the iliopsoas contracture. The piriformis spasm serves to irritate the sciatic nerve. Thus, a left psoas syndrome can cause right gluteal and posterior thigh pain in conjunction with right-sided sciatica.

7. C   Anyone of those three situations can cause the sacral base (top of the sacrum) to move anteriorly, resulting in sacral flexion. Craniosacral extension allows for laxity of the dural membrane (which is connected to the sacrum at S2); such laxity allows the sacrum to fall forward into flexion (nutation in this case) through a mechanism known as inherent motion. Exhalation also allows for laxity throughout the torso, including the dural membrane. That also allows for the sacrum to fall forward into flexion at an S2 horizontal; this mechanism is known as respiratory motion. Torso extension causes the spinal segments to move into extension; the sacrum responds by moving into flexion. Neighboring body regions throughout the body tend to move in opposite directions in order to maintain balance. Accordingly, the sacral flexion that occurs with torso extension helps to maintain balance; this sacral motion is known as postural motion.

8. B   Anytime we are either ambulating or just standing with weight on one leg, we are inducing a sacral torsion on an oblique axis, with the axis ipsilateral to the leg currently bearing the body's weight. This is physiologic as it is a normal component to body movement and positioning. It is only a dysfunction if it gets "stuck" in that position and remains, therefore, in that position.

9. C   L5 right transverse process is deeper on the right than the left, so L5 is rotated left. This is most noticeable in neutral, which means in the freedom in the sagittal plane (flexion/extension/neutral) is found in non-neutral positions (since dysfunction is greatest in neutral). Thus, this is L5 F/E R_LS_L. L5 can induce sacral somatic dysfunction with an ipsilateral sacral oblique axis and contralateral sacral rotation: right sacral rotation on a left sacral oblique axis (R on L).

10. E   All three landmarks of the left innominate are elevated, giving an initial impression of a left superior innominate shear. However, the ASIS compression test (innominate rocking test) is positive on the right, confirming that the pathology is on the right, and not the left. So, the palpatory findings indicated that the left innominate is more superior than the right, but the left innominate is positioned normally and the right is what is in pathology. Thus, it is the right innominate that is pathologically inferior, yielding a right inferior innominate shear. This is consistent with the complaints of left (contralateral) sacroiliac joint pain.

11. A   This is a right anterior innominate rotation. This could be caused by tight quadriceps.

12. C   The standing flexion test is positive on the left, confirming that the pathology is on the left. Thus, if one "perceives" the ASIS to be more lateral on the right, but the real pathology is on the left, then it is the left ASIS that must be pathologically medial. This is, therefore, a left innominate inflare.

13. D   Since the innominate rocking test is positive on the right, the pathology is confirmed to be on the right. The findings lead to a differential diagnosis, therefore, of right superior innominate shear or right superior pubic shear. Thus, the only other findings could be a short right leg and a positive right standing flexion test (for both), and either a superior right PSIS (for a right superior innominate shear) or an inferior right PSIS (for a right superior pubic shear).

14. E   This is a bilateral sacral flexion. Because the sacrum is flexed, there is plenty of soft tissue at the lumbosacral junction to provide for excellent lumbosacral springing during the lumbosacral spring test, yielding a negative lumbosacral spring test. Neighboring body regions move in opposite direction to maintain balance. Thus, as the sacrum moves into flexion, the spine above it moves into extension, causing an increased lumbar lordosis.

15. A   The flexion tests confirm that the pathology is on the left, so we know that, with sacral dysfunction, the left superior sulcus will be pathologically positioned either deep or shallow. Palpation demonstrates it to be deep, with an inferior lateral angle. This indicates that the top left of the sacrum is dipping forward while the bottom right of the sacrum is tipping backwards, causing a torsion around an axis. If the top left and bottom right sacrum are moving like this, they are moving as such around a right axis to create a torsion. Since the left superior sulcus is dipping forward and this is a torsion, this is a forward sacral torsion. Forward sacral torsions always are either right on right, or left on left. Since we already know there is a right axis, we therefore know that this is a right rotation on a right axis (R on R).

16. B   The superior sacral sulcus and the contralateral inferior lateral angle are moving in opposite directions (one shallow, one deep), presenting a situation that appears to be a sacral torsion. The ASIS compression test is positive on the left, confirming that the pathology is on the left. Hence, there cannot be a left axis, since the axis is the most "normally" positioned part of the sacrum and all else is rotating or torsing about it. So, if the pathology is on the left, we know the sacral oblique axis is on the right. That means that it is the left superior sulcus, and not the right, that is moving pathologically. During palpation, it appeared that the right superior sulcus was deeper than the left, but the right is not what is in dysfunction. If this is a torsion and the right superior sulcus is deep but the right is the more normal of the two sulci, then it is the left superior sulcus that is shallow pathologically. Hence, this represents a backward sacral torsion around a right axis. Backward sacral torsions are always either right on left, or left on right. As this has a right axis, we know that it is a left rotation. In conclusion, this is a left on right.

Chapter Six: UPPER EXTREMITIES

THE SHOULDER

Bones:
        1.   clavicles
        2.   humerus
        3.   scapula

Joints:
1. sternoclavicular - the joint between the sternum and the clavicle.
2. acromioclavicular - the joint between the acromion of the scapula and the clavicle.
3. glenohumeral - the joint between the humerus and the glenoid fossa of the scapula.

(plus a "pseudoarthrosis" or "pseudo-joint" created due to the close relationship between the scapula and the posterior thorax, termed the "scapulothoracic joint")

Muscles:

There are many muscles that play a role in stabilization and movement of the shoulder. The primary protection and support of the glenohumeral joint is provided by the **rotator cuff.** The rotator cuff is comprised of 4 muscles:

1. **supraspinatous -** abducts the arm, and is especially involved in the first 90 degrees of abduction
2. **infraspinatous -** externally rotates the arm and is the PRIMARY external rotator (along with teres minor)
3. **teres minor -** externally rotates the arm and is the PRIMARY external rotator (along with the infraspinatous)
3. **subscapularis -** internally rotates the arm; is the PRIMARY internal rotator

These 4 muscles of the rotator cuff may be remembered by the acronym **"SITS".** And, their movements can be remembered as "up, o..u..t, and in". Other muscles of the shoulder include the following:

Pectoralis major - one of 2 primary adductors (the other one is the latissimus dorsi)

Deltoid (anterior portion) - the primary flexor

Deltoid (middle portion) - the primary abductor

Deltoid (posterior portion) - one of three primary extensors (the others are the latissimus dorsi and the teres major)

Teres major - one of three primary extensors (along with the latissimus dorsi and the posterior portion of the deltoid)

Latissimus dorsi - the primary extensor (along with teres major and the posterior portion of the deltoid) and the primary adductor (along with pectoralis major)

<u>**Nerves**</u>

Brachial Plexus: comprised of C5-C8 and T1 nerve roots, and is responsible for innervation of the upper extremity

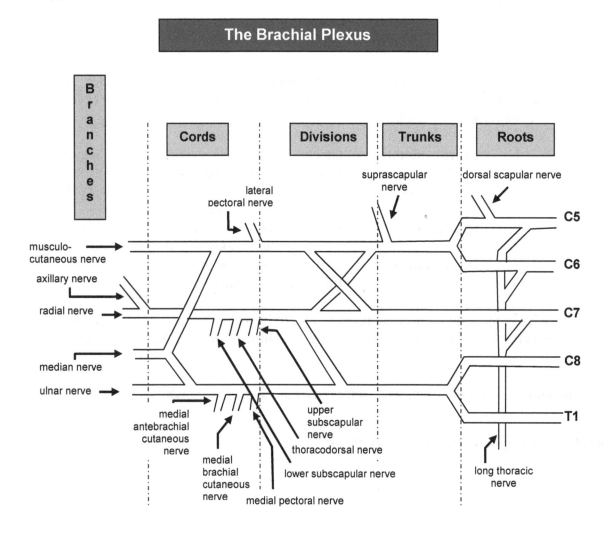

Nerve roots C5 and C6 join to form the upper trunk. The nerve roots of C8 and T1 join to form the lower trunk. C7 remains independent, forming the middle trunk itself. Eventually, the upper trunk and lower trunk divide and contribute to the middle trunk to form the posterior cord. In turn, the middle trunk sends a contribution to the upper trunk to form the lateral cord. Lastly, the lower trunk simply continues laterally to become the medial cord. The terms "medial," "lateral," and "posterior" (when referring to the cords) are used to denote that cord's location in relation to the axillary artery.

The brachial plexus runs between the anterior and middle scalene muscles medially and then runs underneath the clavicle more laterally.

Vasculature

subclavian artery - passes between the anterior and middle scalene muscles.

subclavian vein - passes anterior to the anterior scalene muscle.

axillary artery - the continuation of the subclavian artery, becoming the axillary artery at the lateral border of the first rib

brachial artery - the continuation of the axillary artery, becoming the brachial artery at the inferior border of teres major.

profunda brachial artery - the first major branch of the brachial artery, and travels with the radial nerve in the radial groove of the humerus.

ulnar artery - the brachial artery divides at the bicipital aponeurosis, forming this artery medially; it supplies the elbow, wrist, and hand. In the hand, it forms the superficial arterial arch.

radial artery - the brachial artery divides at the bicipital aponeurosis, forming this artery laterally; it supplies the elbow, wrist, and hand. In the hand, it forms the deep palmar arterial arch.

Lymphatics
The right upper extremity (along with the right side of the head, neck, and thorax) lymphatics drain into the right lymphatic duct while the left upper extremity lymphatics (along with all lymphatics not drained by the right subclavian duct) drain into the thoracic duct.

The thoracic duct drains into the left brachiocephalic vein at the junction of the subclavian and internal jugular veins. The right lymphatic duct drains into the junction of the right internal jugular and right subclavian veins.

The Cervical Nerve Routes versus Arm Sensation

Below is depicted the cervical nerve routes responsible for sensation in the various parts of the arm (as well as shoulder and neck):

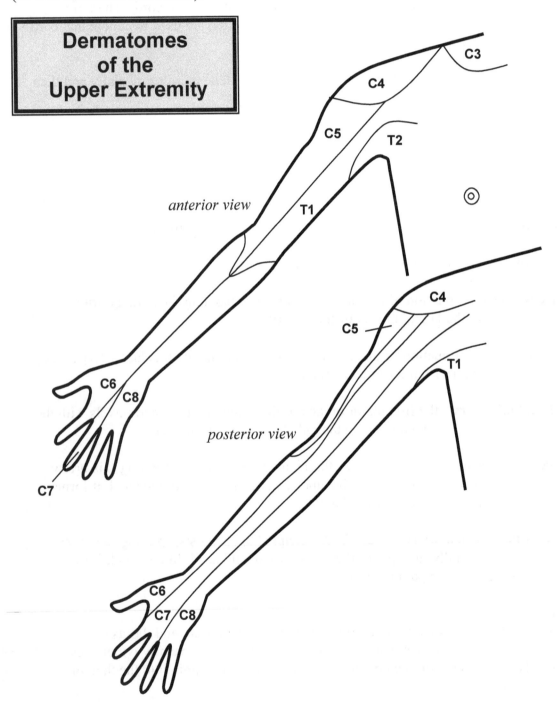

Dermatomes of the Upper Extremity

anterior view

posterior view

Muscles are innervated in the following manner:

Deltoid	C5
Biceps	C5, C6

Brachialis	C5, C6
Wrist extensors (extensor radialis longus, extensor radialis brevis) (but extensor carpi ulnaris is not innervated by C6)	C6, C7
Wrist flexors (flexor carpi radialis, flexor carpi ulnaris, palmaris longus)	C7, C8
Triceps	C7, C8
Interossei (palmar and dorsal)	T1

How does this relate to <u>movements</u> that you can remember?

Arm abduction with elbow flexion (the strongman pose)	C5
Elbow flexion with wrist extension (the "what do you want" pose)	C6
Elbow extension with wrist flexion (the "waiter's tip" pose)	C7, C8
Holding and releasing papers from between fingers	T1

Reflexes to know and the nerve root function they test:

BICEPS REFLEX	**C5**
BRACHIORADIALIS REFLEX	**C6**
TRICEPS REFLEX	**C7**

When testing **deep tendon reflexes (DTRs),** be cognizant that there is a standard method for grading the degree of response:

GRADE	DEGREE OF RESPONSE
4/4	Brisk response with maintained clonus
3/4	Brisk response
2/4	**NORMAL**
1/4	Decreased, but present response
0/4	No response

While testing DTRs, it is often customary to evaluate muscle strength. After all, a general **neurological exam** will always be comprised of tests of MUSCLE STRENGTH, REFLEXES, and SENSATION.

The standard method of recording muscle strength is as follows:

Grade	Diagnosis	Degree of Response
5	**NORMAL**	**movement against gravity *AND* full resistance**
4	Good	movement against gravity and some resistance
3	Fair	movement against gravity but not against resistance
2	Poor	movement *WITHOUT* gravity
1	Trace	some evidence of a contraction, but no movement
0	Zero	no evidence of any contraction

In addition to those tests associated with a neurological evaluation, there are many other tests for evaluating overall motion, in terms of both quantity and quality. There are many tests to evaluate the shoulder:

SHOULDER TESTS

Test	Evaluates/Detects:	Method
Apley's Scratch Test	ROM (range of motion)	Patient reaches behind head to scratch back - evaluates abduction and external rotation; patient reaches across chest to scratch other shoulder and/or reaches around the back at waist and scratches back - evaluates internal rotation and adduction
Adson's Test	Thoracic Outlet Syndrome/its effect on subclavian a.	Patient extends elbow and arm, and slightly abducts arm and, with a deep inhalation, turns head in ipsilateral direction. A (+) test = very decreased or absent radial pulse
Roos Test	Thoracic Outlet Syndrome	Patient abducts both arms to 90 degrees, externally rotates the arms, then flexes the elbows to 90 degrees. The patient repetitively forms a fist and releases it while maintaining the entire arm/forearm position for 3 minutes. A (+) test = much sensation of heaviness or weakness in an arm, or paresthesias of the hand (all on the side ipsilateral to the dysfunction)
Drop Arm Test	Rotator Cuff Tears	Patient abducts arm to 90 degrees, and then slowly and steadily lowers the arm to his/her side. A (+) test = inability to smoothly lower arm and/or arm dropping uncontrollably
Speed's Test	Biceps Tendon	Patient extends elbow and supinates forearm while flexing the arm at the shoulder, while the latter is being resisted by the clinician. A (+) test = tenderness in the bicipital groove (where the biceps tendon is located)
Yergason's Test	Stability of Biceps Tendon in Bicipital Groove	Patient flexes elbow to 90 degrees. The clinician holds the patient's wrist with one hand and the patient's elbow with the other. The clinician then resists the patient's flexive forces while inducing an external rotation of the forearm that the patient resists also. A (+) test = pain in biceps tendon as it pops out of the bicipital groove.

SHOULDER DYSFUNCTION

Dysfunction	Definition	Etiology	Other
Brachial Plexus Injury	injury to brachial plexus, especially during birth	trauma	Several kinds: **Erb-Duchenne**: MOST common; injury to C5-6 nerve roots, resulting in upper arm paralysis **Klumpke's Palsy**: injury to C8-T1 nerve roots, resulting in paralysis of intrinsic muscles of the hand
Winging of Scapula	while patient pushes anteriorly, scapula protrudes posteriorly	long thoracic nerve injury	the nerve injury results in weakness or paralysis of the serratus anterior muscle
Thoracic Outlet Syndrome	an ache and/or paresthesias of neck and/or arm secondary to compression of the neurovascular bundle of the shoulder	cervical rib; spasm or hypertonicity of anterior or middle scalenes; anomaly in pectoralis minor insertion; or, somatic dysfunction of clavicles or upper ribs. Risk factors include poor posture and large/heavy breasts	compression of the neurovascular bundle may occur between the anterior and middle scalenes, and/or the clavicle and 1st rib, and/or the pectoralis minor and the upper ribs (+) Tests = Adson's, Roos, and usually Apley's Scratch. Due to greatly reduced range of motion (ROM), OMT should aid in increasing range of motion while treating the source of the problem. Exercises are also recommended to strengthen the trapezius and levator scapulae.

Dysfunction	Definition	Etiology	Other
Supraspinatous Tendonitis	when the arm is flexed and internally rotated, there is impingement between the acromion and the greater tuberosity	inflammation of the tendon, usually secondary to repetitive action or strain	superior acromion is tender, and is exacerbated by abduction of the arm beyond 60 degrees, with relief increasing as the arm reaches 180 degrees (full vertical) (this pain at the peak of the arc is "painful arc syndrome"). (+) Drop Arm Test, commonly (with pain and inability to slowly lower arm below 90 degrees). Must be "quieted," because continual inflammation will result in calcification. Rest, Ice, Anti-inflammatories, +/- Cortisone injection, and OMT to free restricted ROM are all recommended.
Rotator Cuff Tear	a tear in one of the tendons of the rotator cuff, usually of the supraspinatous	trauma	tenderness just inferior to the tip of the acromion; starts as acute, sharp pain in region followed by ongoing dull ache (especially at night). (+) weakness in abduction and (+) Drop Arm Test. Complete avulsions result in muscle retraction and must be treated surgically. Less severe tears may be treated with rest, ice, NSAIDS, and OMT to free any restrictions in ROM.
Adhesive Capsulitis (Frozen Shoulder Syndrome)	restricted ROM of shoulder with pain, all worsening gradually over time	prolonged immobility or "guarding" of the shoulder, usually secondary to trauma, thoracic outlet syndrome, etc.	tenderness just anterior to the shoulder; more common in patients older than 40. Treatment is prevention! If dysfunction already exists, treat with corticosteroids, and/or NSAIDS, and OMT for freeing restricted ROM and breaking any adhesions.

Dysfunction	Definition	Etiology	Other
Shoulder Dislocation (Subluxation)	humeral dislocation from glenoid	trauma (if repeated, may be sign of psychiatric disorder)	dislocation occurs inferio-anteriorly, often from overextension of the arm. May result in axillary nerve damage, resulting in deltoid muscle paralysis and shoulder anesthesia.
Bicipital Tenosynovitis	inflammation of the tendon of the long head of the biceps, along with its sheath	overuse/ repetitive activities	inflammation leads to development of adhesions that bind the tendon to its sheath. Pain at anterior shoulder (especially over bicipital groove), +/- radiation down arm (all of which is exacerbated by forearm flexion or supination against resistance). Rest, ice, NSAIDS, and OMT to free restricted range of motion (ROM) and tight myofascia is recommended.

The Elbow and Forearm

The elbow is the area created by the articulation of the humerus with the proximal ulna and radius.

Important attachments to know:
Lateral Epicondyle: extensors of wrist and hand - all innervated by the radial nerve
Medial Epicondyle: flexors of wrist and hand - all innervated by the median nerve
(EXCEPTION: flexor carpi ulnaris: ulnar nerve)
The primary extensor of the forearm is the <u>triceps.</u>

The primary flexor of the forearm is the <u>brachialis.</u>

The primary supinators of the forearm are the <u>biceps</u> and the <u>supinator</u>. The former is innervated by the musculocutaneous nerve and the latter by the radial nerve.

The primary pronators of the forearm are the <u>pronator teres</u> and the <u>pronator quadratus,</u> both of which are innervated by the median nerve.

Positioning and Motion of the Elbow and the Forearm

The positioning of the arm with respect to the forearm is measured in terms of the **carrying angle.** The carrying angle is, itself, determined by the degree of abduction of the forearm with respect to the arm, specifically the degree of abduction of the ulna with respect to the humerus. Females have a greater carrying angle then men, meaning that their ulnas normally are more abducted from the humeri than those of men. The **normal female carrying angle is 10-12 degrees,** whereas **that of men is only about 5 degrees.** Anything greater is indicative of somatic dysfunction, and is termed **cubitus valgus.** Conversely, if the angle is less, **cubitus varus** exists and, too, is indicative of somatic dysfunction. The carrying angle is depicted below:

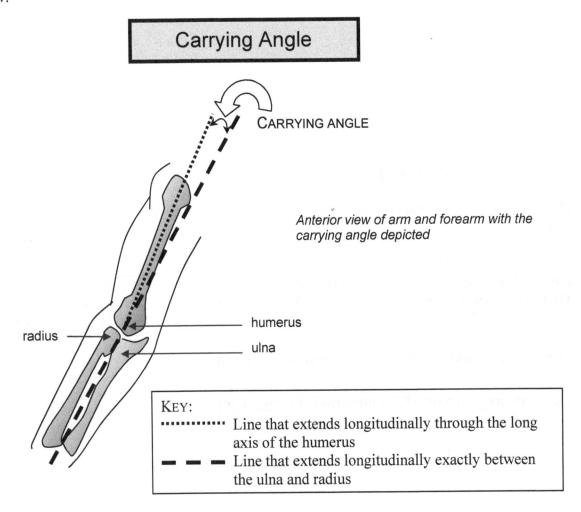

Carrying Angle

CARRYING ANGLE

Anterior view of arm and forearm with the carrying angle depicted

radius

humerus

ulna

KEY:
•••••••••••• Line that extends longitudinally through the long axis of the humerus
━ ━ ━ Line that extends longitudinally exactly between the ulna and radius

The carrying angle has a strong influence on the position of the wrist. In fact, cubitus valgus has the "opposite" effect on the wrist, i.e. if the elbow is in abduction, the wrist is moved into adduction. The same principle applies to cubitus varus.

The radial head also moves, particularly when supinating and pronating of the forearm. When the forearm is pronated, the head glides posteriorly. When the forearm is supinated, the radial head glides anteriorly.

Dysfunction of the Forearm and Elbow

Remember !!!!! All dysfunctions are named for their ease of motion.

Anterior Radial Head: Radial head is "stuck" anteriorly and is restricted in moving posteriorly, meaning that the forearm easily supinates but has restricted pronation.

Posterior Radial head: Radial head is "stuck" in posterior position, meaning that the forearm easily pronates but has restricted supination.

Adduction of the ulna: The ulna is easily adducted, with restriction in abduction. Since, the ulna is in adduction, the wrist will be abducted.

Abduction of the ulna: The ulna is easily abducted, with restriction in adduction. Since the ulna is abducted, the wrist will be adducted.

Lateral Epicondylitis ("tennis elbow"): The extensor muscles are strained at their attachment to the lateral epicondyle. This often results from repetitive strain injury, often involving frequent supination or forceful extension of the wrist with pronation (positions common to tennis). The patient usually has tenderness over or just distal to the lateral epicondyle that tends to worsen with resisted supination; the pain may also radiate along the lateral aspect of the upper extremity. Rest, ice, NSAIDS, and OMT to release myofascial structures and the specific muscles involved and to correct any somatic dysfunction proximal to the site are recommended. Additionally, a several-inch strap is worn around the forearm during activities to reduce the strain by distributing the tension over a greater area.

Medial Epicondylitis ("golfer's elbow"): This results from strain of the flexors as they attach to the medial epicondyle, resulting in tenderness of the medial epicondyle +/- radiating pain over the medial upper extremity. Again, this is usually a result of repetitive strain, especially with strenuous pronation or very forceful flexion of the wrist in combination with supination. Treatment is similar to that of lateral epicondylitis.

The Wrist

Below is depicted the bones of the wrist. One of two ways to remember these bones are to recall "Some Lovers Try Positions That They Can Handle", representing, in order, scaphoid, lunate, triquetrum, pisiform, trapezium, trapezoid, capitate, hamate. Another memory tool to recall the position of the trapezium versus the trapezoid is the use of a rhyming sound: **Thumb - Trapezium**, to remember that it is the trapezium that sits just below the thumb and articulates with the thumb's metacarpal bone.

The Carpal Bones

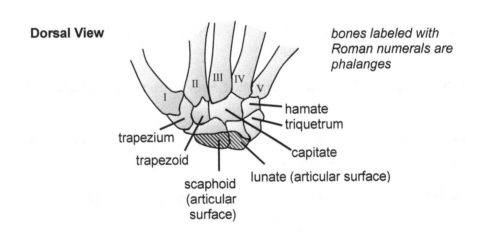

Dorsal View

bones labeled with Roman numerals are phalanges

hamate
triquetrum
capitate
lunate (articular surface)
trapezium
trapezoid
scaphoid (articular surface)

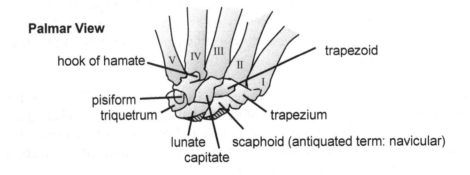

Palmar View

hook of hamate
pisiform
triquetrum
lunate
capitate
trapezoid
trapezium
scaphoid (antiquated term: navicular)

Remember, the carrying angle of the arm directly affects the degree of abduction or adduction of the wrist. Also, when the wrist is in abduction, the radius is moved proximally; alternatively, if the wrist in adduction, the radius moves distally.

Evaluating the Wrist

Test	Evaluates/Detects:	Method
Tinel's	carpal tunnel syndrome	clinician taps over volar aspect of patient's transverse carpal ligament (Tinel⇒ Tap) (+) test = paresthesias of thumb and first 2 ½ fingers
Phalen's	carpal tunnel syndrome	patient's wrist is passively but maximally flexed by the clinician, and is held in this position for one minute. (+) test = paresthesias of thumb and first 2 ½ fingers
Prayer (Reverse Phalen's)	carpal tunnel syndrome	patient grasps clinician's hand, palm to palm, and patient's hand is passively moved into full extension while the clinician places direct pressure onto the carpal tunnel for one minute (+) test = paresthesias of thumb and first 2 ½ fingers
Allen's	radial a. and ulnar a. patency/blood flow (e.g., blood supply to the hand)	patient opens and closes hand several times and then makes a tight fist and holds that position. The clinician occludes the ulnar and radial arteries, and then patient relaxes the hand. The clinician then releases one of the arteries and watches for flushing. The test is failed if the hand does not flush or flushes slowly (indicating that the released artery is not delivering blood adequately). The test is repeated, testing the other of the two arteries.
Finkelstein	tenosynovitis of the abductor pollicis longus and extensor pollicis brevis (DeQuervain's tenosynovitis)	patient makes a tight fist with thumb tucked into the fist. The clinician induces adduction of the wrist. (+) test = pain over the tendons at the wrist

Dysfunction of the Wrist

dysfunction	definition	etiology	other
Carpal Tunnel Syndrome	median nerve entrapment between the longitudinal tendons of the hand and the flexor retinaculum	repetitive strain and activities, particularly with the hand extended	patients have paresthesias of the palmar surface of the thumb and first 2 ½ digits. (+) tests = Phalen's, Prayer, and Tinel's
Drop Wrist Deformity	paralysis of extensor muscles	damage to the radial nerve	patient is unable to extend the wrist, and so wrist tends to "drop" into flexion

The Hand

The interossei and lumbricals of the hand are supplied by C8 and T1 nerve roots. There are seven **interossei:** three palmar and four dorsal. The dorsal interossei abduct the fingers (DAB, as in Dorsal ABduct) and the palmar interossei adduct the fingers (PAD, as in Palmar ADDuct).

There are four **lumbricals,** one for each finger except the thumb. They serve to flex the digits at the metacarpal-phalangeal joints (MCP joints) and to extend the proximal and distal interphalangeal joints (the PIP and DIP joints).

The muscles of the hand are termed the intrinsic muscles of the hand. There are four categories: the thenar muscles, the hypothenar muscles, the interossei and the lumbricals. The thenar muscles are the "thumb muscles" (Thenar → Thumb) and include the abductor pollicis brevis, flexor pollicis brevis, and opponens pollicis which, together, form the thenar eminence. Another thenar muscle is the adductor pollicis muscle; although it is technically a thenar muscle, it does not help to form the thenar eminence and is not innervated by the median nerve. The hypothenar muscles are the "little finger muscles" (hypo means little, as in little finger) and include the abductor digiti minimi, flexor digiti minimi, and opponens digiti minimi. Together they form the hypothenar eminence.

The nerve supply to the hand includes:

Radial Nerve	only provides cutaneous sensory innervation; supplies NO intrinsic hand muscles
Median Nerve	innervates lumbricals 1 and 2, and provides cutaneous sensory innervation to a portion of the hand. The recurrent branch innervates the abductor pollicis brevis, the flexor pollicis brevis, and the opponens pollicis (the thenar muscles of the thenar eminence).
Ulnar Nerve	provides cutaneous sensory innervation to a portion of the hand and innervates ALL muscles of the hand EXCEPT those innervated by the median nerve.

Below are depicted the areas of sensation (cutaneous sensory innervation) provided by the radial, ulnar, and median nerves.

Sensory Nerve Distribution of the Hands

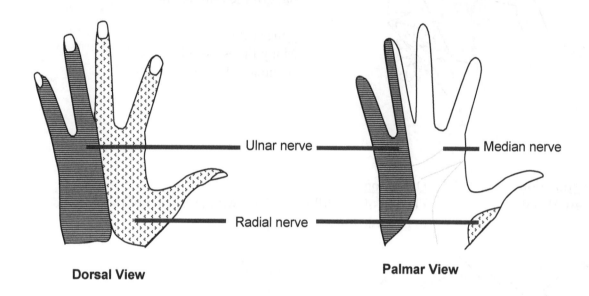

Dorsal View   Palmar View

Dysfunction of the Hand

Dysfunction
Swan neck deformity

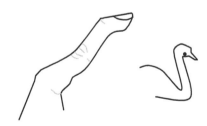

Definition
Flexion contracture of the of the MCP and DIP with extension contracture of the PIP

Etiology
Contracture of intrinsic muscles

Other Info
Often associated with rheumatoid arthritis

Dysfunction
**Boutonniere
deformity**

Definition
Extension contracture of the MCP and DIP with flexion
contracture of the PIP

Etiology
Extension expansion hood rupture of
the extensor digitorum tendon at the
base of the middle phalanx

Other Info
Many times associated with
rheumatoid arthritis

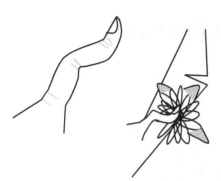

Dysfunction
Claw Hand

Definition
Flexion of the PIPS and DIPS with extension of the MCPs due
to intrinsic muscle activity loss

Etiology
Ulnar nerve damage

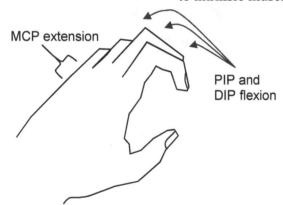

MCP extension

PIP and
DIP flexion

Dysfunction
Ape Hand

Definition
Thenar atrophy (with unopposable thumb)

Etiology
Median nerve damage

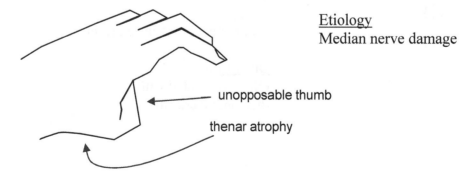

unopposable thumb

thenar atrophy

Dysfunction
Dupuytren's
Contracture

Definition
Flexion contracture of the MCP and PIP with
often contracture of the last two fingers

Etiology
Palmar fascia contracture

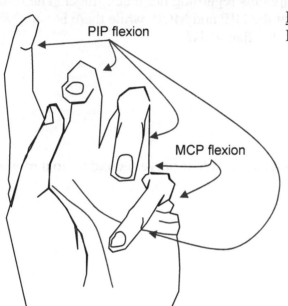

PIP flexion

MCP flexion

Flexion contracture of
the last two digits is
demonstrated, with
extreme contracture of
the fifth digit

Dysfunction
Bishop's
Deformity
(Hand of the Papal
Benediction)

Definition
paralysis of lumbricals 1 & 2

Etiology
Median n. damage

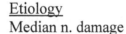

MCP extension of first
and second fingers

NOTE: Some authorities consider Bishop's deformity to
be a deformity that appears much as depicted above, but
that is, instead, associated with contracture of the last two
digits along with atrophy of the hypothenar eminence;
this form of the deformity is due to ulnar nerve damage.

Chapter 6 Review Cases

1. A 59-year old woman presents with complaints regarding her index finger. The finger demonstrates flexion contracture of the DIP and MCP, while there is extension contracture at the PIP. What is the diagnosis?
    a. Boutonniere deformity
    b. Swan neck deformity
    c. Heberden's node
    d. Dupuytren's contracture
    e. Trigger finger

2. A 50-year old female suffers severe damage to her ulnar nerve secondary to a motor vehicle accident. What might result?
    a. Dupuytren's contracture
    b. DeQuervain's tenosynovitis
    c. Ape hand
    d. Claw hand
    e. Drop wrist deformity

3. An 8-year old boy fractures his humerus. The resulting nerve damage may result in what disorder?
    a. DeQuervain's tenosynovitis
    b. Ape hand
    c. Drop wrist deformity
    d. Carpal tunnel syndrome
    e. Ulnar nerve palsy

4. A 42-year old man who has an office position and works continually at his computer now has developed paresthesias on the palmar side of his thumb and first 2 ½ fingers. If this is not addressed and the condition worsens, what might he develop?
    a. Carpal tunnel syndrome
    b. Claw hand
    c. Ape hand
    d. Dupuytren's contracture
    e. Swan neck deformity

5. A teenage athlete is found to have a carrying angle of 15 degrees. How would you describe this patient's status?
    a. Cubitus valgus
    b. Coxa valgus
    c. Cubitus varus
    d. Coxa varus
    e. Halux valgus

6. With respect to the previous patient, what wrist positioning do you expect to find?
   a. Wrist adduction
   b. Wrist abduction
   c. Ulnar adduction
   d. Carpal valgus
   e. Tarsal varus

7. A patient demonstrates a positive Roos test, a positive Adson's test, and an inappropriate response to the Apley's scratch test. What may have caused her condition?
   a. Hypertonicity of the posterior scalenes
   b. Anomaly in pectoralis major insertion
   c. Excellent posture
   d. Long thoracic nerve injury
   e. Cervical rib

8. A 32-year old male suffers cervical injury during a skydiving accident. As a result, he has developed winging of the scapula. What was damaged in the neck to cause this thoracic pathology?
   a. Phrenic nerve
   b. Long thoracic nerve
   c. Dorsal scapular nerve
   d. Suprascapular nerve
   e. Nerve root C8

9. If a patient demonstrates a 2/4 triceps reflex with a ¼ biceps reflex, what intervertebral disc might be herniated if the pathology is due to a herniated disc?
   a. C3
   b. C4
   c. C5
   d. C6
   e. C7

10. During muscle strength testing of the deltoid muscle, it is determined to have 2/5 muscle strength. What may be the cause?
    a. Nothing, as this is a normal muscle strength finding
    b. Musculocutaneous nerve injury
    c. C8 nerve root injury
    d. Axillary nerve injury
    e. Median nerve injury

11. A patient complains of numbness and "strange" sensations along the medial aspect of his forearm. He also complains of sharp neck pain that started abruptly yesterday, shortly before the arm problems began. What tops your list of differential diagnoses?
   a. Thoracic outlet syndrome
   b. Herniation of the C8 intervertebral disc
   c. Compression of T1 nerve root
   d. Herniation of the disc above the T1 segment
   e. Carpal tunnel syndrome

1. B   Flexion contracture at the distal interphalangeal joint and the metacarpal phalangeal joint, along with extension contracture at the PIP, yields the classic swan neck deformity.

2. D   Ulnar nerve damage can cause flexion contracture with intrinsic hand muscle atrophy, demonstrated as flexion of the PIPs and DIPS with extension of the MCPs. Such a positioning results in claw hand.

3. C   A fracture of the humerus can result in radial nerve damage. Radial nerve damage may yield drop wrist deformity due to loss of muscle activity of the extensor muscles.

4. C   The man is currently demonstrating a classic case of carpal tunnel syndrome. Carpal tunnel syndrome results from median nerve entrapment. If he fails to address this problem and the condition worsens, his median nerve may be severely damaged and, over time, result in ape hand.

5. A   A normal carrying angle for females is 10-12 degrees, while the normal carrying angle for males is 5 degrees. The gender of this patient is not noted; however, regardless of the gender, 15 degrees exceeds the normal carrying angle for any person. Thus, this represents a large carrying angle, placing the patient's arm into cubitus valgus.

6. A   The arm is currently angulated outward, causing elbow abduction, as a result to this large carrying angle. Neighboring body regions tend to move in opposite direction in order to maintain balance. Hence, if the elbow is abducted (valgus), then the wrist will be in adduction.

7. E   This patient is demonstrating the classic findings expected for thoracic outlet syndrome. Many things can cause thoracic outlet syndrome, such as hypertonicity of the anterior and/or middle scalenes, anomalies in the pectoralis minor muscle insertion, and, as noted herein, a cervical rib. Poor posture and large breasts can also be risk factors due to the rounded shoulders created by such conditions, causing compression of the neurovascular bundle in the shoulder.

8. B   The long thoracic nerve is formed from cervical nerve roots C5, C6, and C7, and represents the first of the nerves to emanate from the brachial plexus. Cervical injury can damage this nerve, the nerve that innervates the serratus anterior muscle and that is responsible, when injured, for winging of the scapula.

9. B   A 2/4 deep tendon reflex is normal, but a ¼ DTR is lower than normal, representing a lower motor neuron problem. Thus, we know that there is pathology of the nerve root or one of its branches. In this case, the decreased deep tendon reflex was associated with the biceps reflex, and thus with C5. In other words, C5 nerve root is dysfunctional. If this were due to a herniated disc, it would be the C4 intervertebral disc that would be responsible.

10. D   A 2/5 muscle strength test result is poor, indicating severe muscle weakness. Many things can cause muscle weakness. Nerve injury can be one cause. The axillary nerve serves the deltoid muscle, so axillary nerve injury may be the cause of this patient's muscle weakness.

11. D   The abnormal sensations described by this patient are occurring in the C8 dermatome. Such sensations could be due to thoracic outlet syndrome if that had occurred slowly over time and were not associated with an abrupt, acute start in conjunction with acute neck pain. In this patient, sharp neck pain started suddenly the day before, only to be followed soon thereafter by the anomalous arm sensations. Hence, the etiology is more likely to lie with cervical spine pathology. As such, cervical disc herniation is high on the differential. If the C8 nerve root is impinged by a intervertebral disc, it would be the C7 disc that would be responsible. The C7 disc is the disc that lies above or "on" the T1 vertebral segment, and that exists beneath the C7 vertebral segment.

Chapter Seven: LOWER EXTREMITIES

The Leg

<u>Anatomy</u>

BONES:
Femur
Tibia
Fibula
Patella   a sesamoid bone that attaches to the quadriceps tendon and the patellar tendon

<u>Muscles</u>
GLUTEUS MAXIMUS   primary hip extensor

ILIOPSOAS   primary hip flexor

QUADRICEPS   primary knee extensor

SEMIMEMBRANOSIS & SEMITENDINOSIS   primary knee flexors

"Q u a d s " = quadriceps, a group of muscles comprised of the rectus femoris, vastus lateralis, vastus medialis, and vastus intermedius muscles.

"H a m s t r i n g s " = the posterior thigh muscles, and include the semitendinosus, semimembranosis, and biceps femoris muscles.

<u>Joints</u>

Hip Joint = **Femoroacetabular Joint** (a ball and socket joint)
Knee joint => composed of three joints:
                    1.   **TIBIOFEMORAL JOINT**
- the largest joint in the body
- = the tibial and femoral articulation
- in actuality, the tibia and femur are physically separated from one another by the medial and lateral menisci
(the menisci serve to aid the joint in nutrition, lubrication, and shock absorption)

2.   **PATELLOFEMORAL JOINT**
3.   **TIBIOFIBULAR JOINT**

<u>Ligaments</u>

The femoroacetabular joint is supported by 4 ligaments:
1.   **Iliofemoral Ligament**
2   **Ischiofemoral ligament**
3.   **Pubofemoral ligament**
4.   **Capitis femoris**

The Femoroacetabular Joint

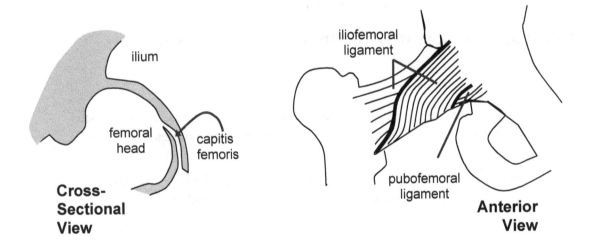

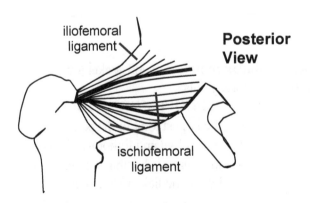

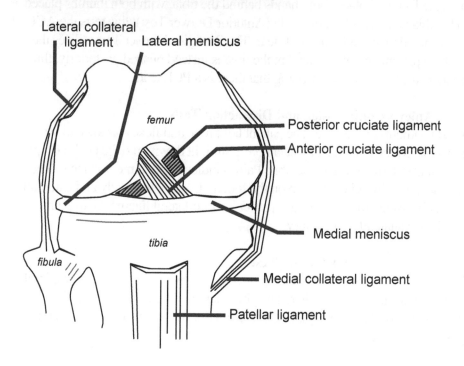

The Tibiofemoral Joint

An anterior view of the right knee with the patella removed

Lateral collateral ligament
Lateral meniscus
femur
Posterior cruciate ligament
Anterior cruciate ligament
Medial meniscus
tibia
fibula
Medial collateral ligament
Patellar ligament

The knee joint is supported by 4 ligaments:

1. **Anterior cruciate ligament (ACL)**
    - connects the anterior tibia to the posterior femur (prevents knee hyperextension)
2. **Posterior cruciate ligament (PCL)**
    - connects the posterior tibia to the anterior femur (prevents knee hyperflexion)
3. **Lateral Collateral Ligament (LCL; Fibular Collateral Ligament)**
    - connects the femoral lateral epicondyle to the lateral surface of the fibular head (prevents disruption of the sides of the knee joint)
4. **Medial Collateral Ligament (MCL; Tibial Collateral Ligament)**
    - connects the medial epicondyle of the femur to the medial condyle and superior aspect of the medial surface of the tibia (prevents disruption of the sides of the knee joint)
    - is physically connected to the medial meniscus

Tests to Assess the Knee:

Anterior and Posterior Drawer Tests

This is used to assess the ACL and PCL. The patient lies supine with one knee flexed. The physician sits on the patient's foot of the flexed leg and places both hands behind the tibia, with both thumbs placed at the joint line bilaterally. The tibia is then pulled anteriorly (Anterior Drawer Test); this tests the ACL. If there is laxity, the test is considered positive for an ACL tear. The Posterior Drawer Test utilizes the same positions on behalf of the patient and physician, but the tibia is instead pushed posteriorly, thus testing the PCL. Laxity renders the test positive, meaning that there is a PCL tear.

Apley's Compression and Distraction Tests

These assess the menisci and ligaments of the knee. The patient lies prone and flexes his knee to 90 degrees. The Compression Test requires that the physician press straight down onto the heel (i.e. apply compression) plus internally and externally rotate the tibia. Pain is indicative of a meniscal tear. Then, the distraction test is performed: the physician pulls up on the foot and, again, internally and externally rotates the tibia. Pain is indicative of ligamentous injury, particularly collateral ligaments. In both situations, the side of the pain is ipsilateral to the injury.

Lachman's Test

This evaluates the stability of the ACL. The patient lies supine and the physician grasps the proximal tibia with one hand and the distal femur with the other. The physician then flexes the knee to 30 degrees, and then pulls the tibia forward (anteriorly). Excessive anterior movement of the tibia renders the test positive, and indicates an ACL tear.

McMurray's Test

This test identifies the existence of tears of the posterior aspect of the menisci. The patient assumes the prone or seated position. The lateral meniscus is evaluated by the physician by fully flexing the knee while also palpating the lateral aspect of the joint line. The tibia is then internally rotated and the knee is varus stressed. While maintaining this position, the leg is slowly extended as the lateral knee is palpated. The test is positive if a click is appreciated. The medial meniscus is tested in a similar manner, except that the medial joint line is palpated while EXTERNAL rotation is exerted, and a valgus stress in applied, and then the slow extension of the leg is induced. A memory tool for remembering leg rotation direction with varus or valgus stress application is as follows: the patient's toes always will be pointed in the direction of the valgus/varus stressed position.

Patellar Femoral Grinding Test

This is used to assess the integrity of the posterior patellar surface and the trochlear groove of the femur. The patient lies supine with the legs relaxed. The clinician slides the patella distally, and then instructs the patient to flex his quadriceps while the clinician offers some resistance to the patella. The patellar movement should be smooth and gliding, free of palpable crepitation. Crepitance renders the test positive, and is often accompanied by pain or discomfort. A positive test is found in such disorders as chondromalacia patellae, osteochondral defects, or degenerative changes within the trochlear groove itself.

Crepitation = a grating sound or sensation

Nerves

Femoral nerve (L2-L4) motor: quadriceps, iliacus, sartorius, pectineus
 sensory: anterior thigh and medial leg

Sciatic nerve (L4-S3) = the main branch of the sacral plexus and is the largest nerve in the body
 2 Divisions:
 1. **Tibial Division of the Sciatic Nerve**
  ⇒ motor: hamstrings (EXCEPT short head of biceps femoris), most plantar flexors, toe flexors, and foot invertors
  ⇒ sensory: lower leg and plantar foot
 2. **Peroneal (Common Fibular) Division of the Sciatic Nerve**
  ⇒ motor: short head of biceps femoris, evertors and invertors and dorsiflexors of foot, and most toe extensors.
  ⇒ sensory: lower leg and dorsum of foot

The tibial division of the sciatic nerve becomes the medial and lateral plantar nerves which supply the intrinsic muscles of the foot except for the extensor digitorum brevis (which, itself, is innervated by the peroneal nerve).

Neuro Exam – remember, a properly conducted neuro exam ALWAYS includes muscle strength, sensation, and reflex testing!!

NERVE ROOTS: MUSCLE INNERVATION, REFLEXES, AND SENSORY SUPPLY

Nerve Root	Muscle Innervation	Reflex	Sensation
L1	Iliopsoas	N/A	Upper anterior thigh
L2	iliopsoas, quadriceps, adductors	N/A	middle anterior thigh
L3	iliopsoas, quadriceps, adductors	N/A	lower anterior thigh
L4	anterior tibialis, quadriceps	Patellar	medial malleolus
L5	extensor hallucis longus	N/A	dorsum of foot
S1	peroneus longus and brevis, gastrocnemius	Achilles	lateral malleolus

Correlating Muscles, Nerve Roots, and Movement

Muscle	Movement	Nerve Root
Iliopsoas	Hip flexion (the primary one!)	L1, L2, L3
Gluteus maximus	Hip extension	(L5), S1, S2
Hamstrings	Hip extension	L5, S1, (S2)
Adductors	Thigh adduction (the adductors include the gracilis, pectineus, adductor longus, adductor brevis, and adductor magnus muscles)	L2, L3
Gluteus medius & minimus	Thigh abduction	L5, S1
Piriformis	Abduction of flexed thigh	S1, S2
Quadriceps	Extension of leg at knee	L2, L3, L4
Hamstrings	Leg flexion at knee	L5, S1, (S2)
Anterior tibialis	Dorsiflexion & inversion of foot	L4
Extensor hallucis longus	Foot dorsiflexion and great toe extension	L5
Gastrocnemius	Foot plantar flexion	S1
Peroneus longus & brevis	Foot eversion	S1

Dermatomes of the Lower Extremity

Anterior View

Posterior View

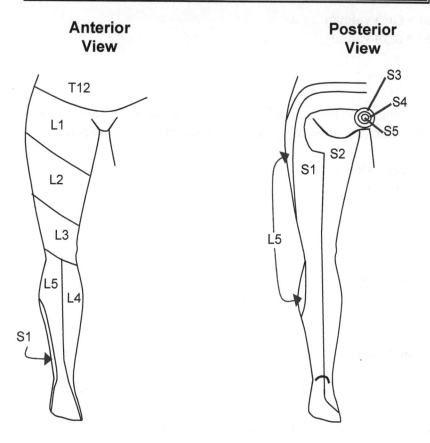

Note: The dorsal foot and toes are innervated by L5. Only the most medial aspect of the great toe and dorsal foot (e.g., the inside side of the great toe and foot) is innervated by L4. Likewise, only the most lateral aspect of the digiti minimi and dorsal foot is innervated by S1. The bulk of the plantar foot is innervated by S1, with the exception of a small area of the medial arch and calcaneal area that is innervated by S2.

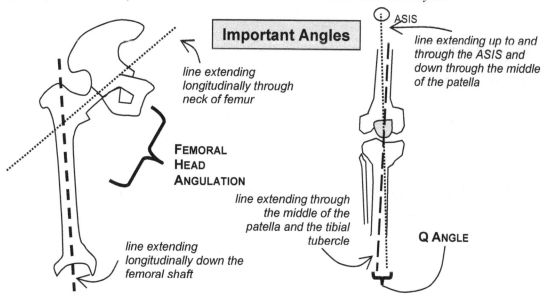

Important Angles

line extending longitudinally through neck of femur

FEMORAL HEAD ANGULATION

line extending longitudinally down the femoral shaft

ASIS

line extending up to and through the ASIS and down through the middle of the patella

line extending through the middle of the patella and the tibial tubercle

Q ANGLE

Femoral head angulation - the angulation between the neck of the femur and the shaft of the femur.

\> 135 degrees = coxa valgum
\< 120 degrees = coxa varum

Q Angle - the angulation between a line drawn from the ASIS through the middle of the patella and a line from the tibial tubercle through the middle of the patella.

\>12 degrees = genu valgum
\<10 degrees = genu varum

Movement Considerations

Facts to Know:

Pronation of the foot/ankle = dorsiflexion, eversion, and abduction of the foot

Supination of the foot/ankle = plantar flexion, inversion, and adduction of the foot

Pronation of the foot causes the fibular head to move anteriorly.

Supination of the foot causes the fibular head to move posteriorly.

Somatic Dysfunction

FIBULAR HEAD DYSFUNCTION - noted by restricted motion, leading often to knee pain.

Posterior fibular head: restricted anterior glide of the fibular head plus the foot appears more supinated (this may affect the peroneal nerve, too, because this nerve rests posterior to the fibular head!!!)

Anterior fibular head: restricted posterior glide of the fibular head plus the foot appears more pronated.

Compartment Syndrome

This is marked by an increase in intracompartmental pressure in the lower leg; it usually results from trauma or severe overuse of the leg. There are 4 compartments within the lower leg, each of which can accumulate pressure to yield a compartment syndrome:

1. Lateral — due to hyperpronation of the foot; pain is posterioinferior to lateral malleolus
2. Superficial posterior — produce the "common shin splints"
3. Deep posterior — pain medial and slightly anterior to the mid-tibia. It occurs most frequently in novice runners, and is due to strain of the posterior tibial tendon, the flexor digitorum longus, and the flexor hallucis longus muscles.
4. Anterior — the MOST common one, characterized by a hard, tender anterior tibialis. It produces the "lateral shin splints", and is characterized by lateral and slightly anterior mid-tibial pain. It may be caused by running up hills or running on toes. It is due to stress on the anterior tibialis tendon, the extensor digitorum longus, and the extensor hallucis longus muscles.

If the pressure becomes too great, the arterial perfusion may be decreased, necessitating a surgical fasciotomy. The latter is generally reserved for severe cases of compartment syndrome that can occur after trauma, especially cases that cause accumulation of blood or excessive edema in any of the compartments.

Muscle Injury = STRAIN (but can also involve distortion of ligamentous structures)

Ligamentous Injury = SPRAIN

There are three sprain grades:

Grade	Tear	Tensile Strength	Laxity
First degree	0	+	0
Second degree	Partial	Decreased	+
Third degree	Complete	0	++++ (severe)

A third degree sprain may require surgical intervention.

Lateral Femoral Patellar Tracking Syndrome

This syndrome is associated with a strong vastus lateralis and weak vastus medialis, resulting from a wide Q angle such as that often found in women. This ultimately results in lateral deviation of the patella that eventually causes irregular or more rapid wear of the patellar posterior surface. Patents will exhibit deep knee pain that is exacerbated by climbing stairs. Treatment is geared towards exercises to strengthen the vastus medialis such as extending the leg against resistance.

O'Donahue's Triad

This is the **Terrible Triad.** It is an extremely common knee injury, and involves the ACL, MCL (medial collateral ligament), and medial meniscus. It results from being struck on the lateral side of the knee (valgus stressed).

Popliteal (Baker's) Cyst

The "cyst" is actually an enlargement of the semimembranosis bursa. In adults, it may be a result of a meniscal tear, rheumatoid arthritis, or other joint dysfunction. It is located lateral to the medial hamstring in the popliteal fossa.

Osgood-Schlatter Disease

This disease involves the tibial tuberosity, and is most common in those who are 11-15 years old. It is accompanied by pain and swelling over the tibial tuberosity, and is exacerbated by squatting, climbing stairs, and extending the knee against resistance. Radiographically, the tibial tuberosity may appear separated with new bone growth beneath it. Treatment is primarily geared towards modifying physical activity to decrease stress on the tendon. More severe cases require a knee splint or cast.

Chondromalacia patellae

This disorder is characterized by softening and fraying of the patellar cartilage (which is on the posterior surface of the patella). It causes anterior knee pain, is usually bilateral, and is usually exacerbated by climbing hills or stairs. Treatment involves NSAIDs and activity modification.

Ankle and Foot

Anatomy

THE BONES OF THE FOOT

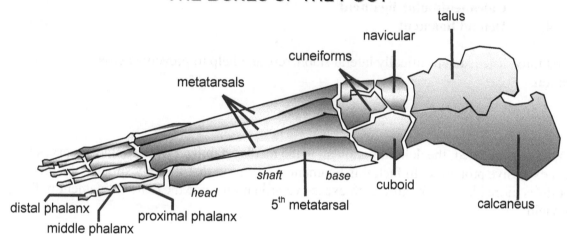

Joints

There are 30 synovial joints in the foot and ankle. The two more important joints are:

1. **Talocrural joint (tibiotalar joint)**
   - a hinge joint that is the articulation between the medial malleolus and the talus
   - the main motion of this joint is plantar flexion and dorsiflexion
2. **Subtalar joint (talocalcaneal joint)**
   - the articulation between the talus and the calcaneus
   - the joint has several functions: it acts as a shock absorber and also allows internal and external rotation of the leg while the foot is planted on the floor.

Ligaments

There are over 100 ligaments in this region of the body. However, the primary ligamentous stabilizers are:

1.   **Anterior talofibular ligament**
2.   **Posterior talofibular ligament**
3.   **Calcaneofibular ligament**
4.   **Deltoid ligament**

The first three listed are specifically lateral stabilizers and help to prevent excessive supination.

The last ligament listed, the deltoid ligament, is the medial stabilizer of the ankle. It serves to prevent excessive pronation. In fact, this ligament is so strong that it rarely sprains; the medial malleolus is more likely to fracture with excessive pronation since the deltoid ligament will rarely yield.

Two other important ligaments are the **Plantar Ligaments:**
1.   **Spring Ligament (Calcaneonavicular ligament)**
     o this aids the medial longitudinal arch
2.   **Plantar Aponeurosis**
     o this is a strong connective tissue connection that stretches from the calcaneus to the phalanges.

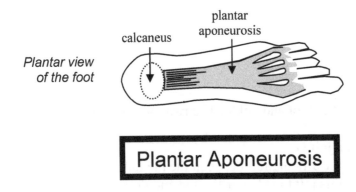

Plantar view of the foot

calcaneus

plantar aponeurosis

Plantar Aponeurosis

Arches

There are 2 classifications of arches in the foot
1.   *Longitudinal Arches*
2.   *Transverse Arches*

The longitudinal arches are comprised of the medial and lateral longitudinal arches. The medial longitudinal arch involves the first thee metatarsals, the cuneiforms, the navicular, and the talus. The lateral longitudinal arch involves the 4th and 5th metatarsals, the cuboid, and the calcaneus.

The transverse arch involves the navicular, the cuneiforms, and the cuboid.

MOTION

FACTS TO KNOW

The ankle is more stable in dorsiflexion because of the configuration of the ankle mortise (superior portion of the talus).

Chronic irritation or inflammation of the plantar aponeurosis can cause calcium deposition, resulting in a heel spur.

Sprains are common to the ankle. The lateral ligaments (posterior talofibular, calcaneofibular, and anterior talofibular) are the most commonly involved, and are associated with oversupination. Of these three ligaments, it is the anterior talofibular ligament that is most frequently injured. There is a special classification system for the sprains of these three ligaments:

ANKLE SPRAIN CLASSIFICATIONS

Type I	sprained anterior talofibular ligament
Type II	sprained anterior talofibular and calcaneofibular ligaments
Type III	sprained anterior talofibular, calcaneofibular, and posterior talofibular ligaments

Somatic Dysfunction of the Arches of the Foot

Of all of the arches of the foot, the one most subject to somatic dysfunction is the transverse arch. The usual cause is displacement of the navicular, cuboid, and/or cuneiforms. This is a painful condition, common to people who are actively on their feet for long periods of strenuous time (such as marathon runners). Generally, if it is the cuboid that is involved, its medial edge tends to move towards the sole. Navicular involvement is such that its lateral side moves toward the sole. And, lastly, cuneiform involvement is most common to the 2nd cuneiform whereupon it glides toward the sole.

Chapter 7 Review Cases

1. A 19-year old college football player severely valgus stresses his knee during a game. What is the most likely injury he has suffered as a result to this?
   a. Lateral collateral ligament damage with a tear to the lateral meniscus
   b. Tear to the posterior cruciate ligament
   c. Tibial collateral ligament damage with a tear to the medial meniscus
   d. Lateral femoral patellar tracking syndrome
   e. Tear to the patellar ligament

2. A 50-year old male is evaluated for complaints of right knee pain. The Apley's compression test is positive on the lateral side of the right knee, and the McMurray's test is positive when the right knee is internally rotated and varus stressed. What is the diagnosis?
   a. Fibular collateral ligament damage
   b. Medial meniscus damage
   c. Medial collateral ligament damage
   d. Lateral meniscus damage
   e. Chondromalacia patellae

3. A child is brought by his parents due to the child's complaints of "pins and needles" sensations on the top of his foot, along with difficulty dorsiflexing the foot. What may be the cause?
   a. Osteoarthritis of the foot
   b. Third degree, Type II sprain of the ankle
   c. Anterior compartment syndrome
   d. Anterior fibular head
   e. Fibular head with restricted anterior glide

4. A patient is referred to your for an L5 herniated nucleus pulposus. What pathology do you expect to find on examination?
   a. Paresthesias of the lateral superior lower leg in the area just under the knee, lack of sensation on most of the dorsum of the foot, and weakness of gluteus medius and minimus
   b. Paresthesias of the lateral ankle and lateral aspect of the foot, some weakness of the iliopsoas, and an increased Achilles reflex
   c. Paresthesias of the lateral ankle, lateral plantar foot, and posterior leg from buttocks to ankle, along with weakness of the gastrocnemius and hamstrings, plus a decreased Achilles reflex
   d. Paresthesias of the medial leg below the knee and the most medial aspect of the foot, weakness of the anterior tibialis, and a decreased patellar reflex
   e. Paresthesias of the area immediately surrounding the anus, and weakness of the gluteus maximus

5. A large femoral head angulation has been identified on an elderly patient. In what does this result?
    a. Coxa valgus with genu valgus
    b. Coxa valgus with genu varus
    c. Coxa varus with genu valgus
    d. Coxa varus with genu varus
    e. Coxa varus with genu varus and tarsal varus

6. A 13-year old boy who has just started with a soccer team is now reporting achy pain at the top center of his "shin bone." He says it gets particularly worse when they are told to run up the stadium steps for exercise and to do squats. What is the likely diagnosis?
    a. Anterior compartment syndrome
    b. Posterior fibular head
    c. O'Donahue's triad
    d. Osgood-Schlatter disease
    e. Chondromalacia patellae

7. A 7-year old boy is suffering from 3- months of low grade hip pain that started slowly and insidiously. He complains of pain in the inguinal area that is exacerbated with rotation of the thigh. Rest improves the condition. The child is not obese, and has no history of trauma. Femoral head sclerosis is demonstrated on frog-leg A-P pelvic hip X-ray. What is the diagnosis?
    a. Chondromalacia patellae
    b. Osgood-Schlatter disease
    c. Legg-Calve-Perthes disease
    d. Slipped capitis femoris epiphysis
    e. Terrible triad

8. A young woman trips and sprains her ankle on the walking trail at the local park. Examination demonstrates decreased tensile strength as well as increased laxity of the ankle, with evident damage to the anterior and posterior talofibular ligaments as well as the calcaneofibular ligament. How would you describe the sprain?
    a. First degree, Type I ankle sprain
    b. First degree, Type II ankle sprain
    c. Second degree, Type II ankle sprain
    d. Second degree, Type III ankle sprain
    e. Third degree, Type II ankle sprain

Answers to Chapter 7 Cases

1. C   Valgus stress of the knee means that the distal tibia moves more lateral than normal, placing strain on the medial structures of the knee. This stretches and potentially damages the medical collateral ligament, also known as the tibial collateral ligament. Because the medial collateral ligament is attached to the medial meniscus, any such stretching or damage to the MCL tends to also cause tearing of the medial meniscus.

2. D   The Apley's compression test evaluates the medial and lateral menisci of the knee; a positive finding on the lateral side of the knee indicates that there is damage to the lateral meniscus. McMurray's test is also used to evaluate the menisci of the knee, with particular focus on the posterior aspect of those menisci. A positive finding with this test when the leg is varus stressed and internally rotated at the knee indicates that there is damage to the posterior aspect of the lateral meniscus. Thus, this is clearly a lateral meniscus pathology.

3. E   The dorsum of the foot is supplied by the peroneal (common fibular) division of the sciatic nerve. The same division also supplies the muscles of the foot that control dorsiflexion. Thus, this appears to be a condition involving impairment of the peroneal nerve (peroneal division of the sciatic nerve). One of the more common causes for peroneal nerve impairment is a posterior fibular head. If there is a posterior fibular head, that means the fibular head's freedom of movement or preferred position is posterior; if a posterior position represents the freedom (its ease of motion), then the restriction is in anterior glide. Thus, if this child has restriction in anterior glide of the fibular head, he could demonstrate the problems about which he complains.

4. C   L5 disc herniation, if it causes nerve root impingement, will first impinge the S1 nerve root. S1 supplies many dermatomes, including those involving the lateral ankle, lateral plantar foot, and a strip of the posterior leg extending from buttocks to ankle; paresthesias could result in those locations. S1 also serves a number of muscles, including the gastrocnemius and the hamstrings; weakness would result if S1 were impinged. Lastly, S1 can be evaluated by the Achilles reflex. A decreased reflex, rather than an increased reflex, indicates that the pathology stems from a nerve root or any of its more distal branches. Thus, having a decreased deep tendon reflex (DTR) reflects injury to a nerve root (or distal branch); since it is the Achilles reflex that is anomalous, we identify S1 as the culprit, which would be consistent with an L5 disc herniation.

5. B   A large femoral head angulation indicates that there is coxa valgus (the distal femur is more lateral than normal, yielding a large femoral head angulation). Since neighboring body regions tend to move in opposite directions to maintain balance, a coxa valgus would generally result in a genu varus (varus at the knee; distal tibia more medial than normal).

6. D   The patient complains of pain at the top center of his tibia (shin bone), the location of the tibial tuberosity. He has recently become athletically active and is between the ages of 11-15, raising the suspicion of Osgood-Schlatter disease. He admits to worsening of pain when climbing steps and when squatting, two activities that put strain on the patellar ligament and its attachment, the tibial tuberosity. This demonstrates the distinct features of Osgood-Schlatter disease.

7. C   There is a several month history of insidious-onset hip pain in a pediatric patient; this raises the suspicion of Legg-Calve-Perthes disease or slipped capitis femoris epiphysis. It is confirmed to be a hip pathology since pain is exacerbated upon rotation of the thigh at the hip, and that pain is located to the inguinal area (this is a finding for <u>any</u> true hip joint (femoroacetabular joint) pathology). The fact that the patient is not obese and has had no trauma history lessens the likelihood of slipped capitis femoris epiphysis. However, the X-ray findings confirm the diagnosis of Legg-Calve-Perthes disease by demonstrating femoral head sclerosis, best seen on frog-leg A-P hip X-ray. A slipped capitis femoris epiphysis would demonstrate, instead, a slipped capitis femoris epiphysis, appearing much like an "ice cream scoop falling off a cone" in the area of the femoral head.

8. D   The patient's ankle demonstrates decreased tensile strength and increased laxity, indicating that this is a second degree sprain. Since it is an ankle sprain, the type of sprain is also determined, based upon what lateral ankle ligaments are involved. All three are involved, making this a Type III ankle sprain. Hence, this is a second degree, Type III ankle sprain.

Chapter Eight:
INNERVATION AND NEURAL CONNECTIONS

A reflex is the response to a stimulus that arises in a receptor/receptive neuron and passes to a nerve center and then outward to an effector such as a gland or muscle. The typical spinal reflex, thus, is comprised of an afferent limb representing the sensory input, a central limb representing the spinal pathway, and an efferent limb representing the motor pathway to muscles or the autonomic nervous system to viscera. The sensory input that arrives via the afferent limb can actually arise from the brain, viscera (via autonomic visceral afferents), or from somatic afferents transmitting information from muscle spindles, nociceptors, Golgi receptors, and so forth.

The 6 types of reflexes are:

1. **VISCERO-SOMATIC**
    Localized visceral stimuli produce a reflex response via same-segment somatic efferents. For example, localized stimulus of an organ may produce a response or tenderpoint in a muscle or other somatic structure whose innervation arises from the same segment.

2. **SOMATO-VISCERAL**
    Localized somatic stimuli produce a reflex response via same-segment visceral efferents. For example, a tenderpoint over a particular muscle may produce a response in the viscera whose innervation arises from the same segment.

3. **PSYCHO-VISCERAL**
    Psychic stimuli such as stress, anxiety, depression, etc. may produce a reflex response in various organs whose innervation is received, at least in part, from the brain (i.e., via cranial nerves such as the vagus). Alternatively, the psychic stimuli may work through the neuroendocrine axis, effecting profound visceral changes via endocrine control. Furthermore, stress and anxiety provoke sympathetic nervous system stimulation, causing multiple visceral changes. In extreme cases of sympathetic nervous system stimulation, the adrenal medulla releases volumes of epinephrine and norepinephrine into the bloodstream, causing systemic rather than segmental sympathetic effects.

4.   **PSYCHOSOMATIC**
Psychic stimuli such as those noted above produce a reflex response in somatic structures whose innervation is received, at least in part, from the brain (i.e., cranial nerves such as the facial nerve and the trigeminal nerve). A psychosomatic *reflex* must involve the reflex components. Alternatively, an indirect effect may be realized by mental states such as anxiety and stress that ultimately stimulate the sympathetic nervous system, causing significant changes in somatic structures. In extreme stimulation of the sympathetic nervous system, the adrenal gland may be stimulated to release epinephrine; the vascular and metabolic changes incurred in this fashion may have striking effects on the somatic structures.

5.   **SOMATO-SOMATIC**
Localized somatic stimuli produce a reflex response in somatic structures whose innervation is derived from the same segment. For example, a strained muscle may produce muscle dysfunction in other muscles innervated by the same segment

6.   **VISCERO-VISCERAL**
A localized visceral stimulus produces a reflex response in visceral structures whose innervation is derived from the same segment.

Within the context of understanding reflexes, it is therefore important to know the segmental sympathetic and parasympathetic innervation of the major organs. This facet of OMT is one of the most important features to know, so do commit these innervations to memory!

Sympathetic Innervation

Viscera	Spinal Cord Level	Presynaptic Sympathetic Fiber	Sympathetic Collateral Ganglion
Head	T1-T4		
Neck	T1-T4		
Heart	**T1-T6**		
Respiratory System	**T1-T6**		
Esophagus	T1-T6		
Upper GI:	**T5-T9**	**greater splanchnic nerve**	**celiac ganglion**
spleen			
stomach			
liver			
gallbladder			
part of pancreas and duodenum			
Arm structures	T2-T8		
Middle GI:	**T10-T11**	**lesser splanchnic nerve**	**superior mesenteric ganglion**
jejunum			
ileum			
ascending colon			
proximal 2/3 of transverse colon			
part of the pancreas and duodenum			
Lower GI:	**T12-L2**	**least splanchnic nerve**	**inferior mesenteric ganglion**
descending colon			
sigmoid colon			
rectum			
distal 1/3 of transverse colon			

Adrenal medulla	T10	**Leg structures**	T11-L2
Kidneys	**T10-T11**	Uterus & Cervix	T12-L2
Upper ureters	T10- T11	Clitoral erectile tissue	L2
Lower ureters	T12-L1	Penile erectile tissue	L2
Bladder	T12-L2	Appendix	T12
Testes & Ovaries	**T10-T11**		

Please be aware that every effort has been made to provide the most widely accepted and most 'board friendly' innervations. However, there is substantial variation within the literature. It is suggested that you learn one set of innervations, rather than trying to learn all published variations. You will usually be able to answer any board questions posed and handle any patient-related clinical problems if you know one set thoroughly.

Parasympathetic Innervation

Viscera	*Spinal Cord or Cranial Nerve Level*
Pupils	Cranial nerve (CN) III via ciliary ganglion
Lacrimal and Nasal Glands	CN VII via pterygopalatine ganglion
Submandibular and Sublingual Glands	CN VII via submandibular ganglion
Parotid Glands	CN IX via otic ganglion
All thoracic organs (including inferior 2/3 of the esophagus), all abdominal organs up to the distal 1/3 of transverse colon, kidney, superior 1/2 of ureters, and testes & ovaries	brain through C0 via vagus nerve (CN X)
distal 1/3 of transverse colon and remaining alimentary canal, inferior 1/2 of ureters, all pelvic organs except ovaries and testes	S2-S4 via pelvic splanchnic nerve

One should also be very familiar with what the visceral effect is if the sympathetic versus parasympathetic nervous system is stimulated:

Viscera	Sympathetic Response	Parasympathetic Response
Pupils	Mydriasis	Miosis
Lens	Ciliary muscle relaxes	Ciliary muscle contracts, allowing lens to become convex for near vision
Bronchi	Dilation	Constriction

Viscera	Sympathetic Response	Parasympathetic Response
Heart	Increased heart rate and force	Decreased heart rate and force
Coronary arteries	Dilated (via beta-2 receptors) or constricted (via alpha-1 receptors); dilation is the dominant response	Dilated
Intestines	Decreased motility and secretions	Increased motility and secretions
GI sphincters	Constriction	Relaxation
Rectum	Filling allowed	Emptied
Bladder wall (detrusor)	Relaxation	Contraction
Bladder sphincter (trigone)	Contraction	Relaxation
Penis	Ejaculation	Erection
Remember: Point (Parasympathetic) & Shoot (Sympathetic)		
Salivary glands	Vasoconstriction, thus slight but thick secretions	Copious, watery secretions
Adrenal medulla	Secretion of catecholamines	None
Liver	Glucose released (secondary to gluconeogenesis and glycogenolysis)	Slightly increased glycogenesis
Gallbladder	Relaxation	Contraction
Uterus	Relaxation	Contraction
Arteries	Constriction in skin and viscera (via alpha-1 receptors) and relaxation in muscle (via beta-2 receptors)	Little or none
Kidneys	Decreased output	None directly
Skeletal muscle	Increased glycogenolysis and strength	None directly

Chapter 8 Review Cases

1. A patient suffers from excess perspiration of his left lower torso and upper anterior thigh whenever he has exacerbation of his chronic rectal spasms. What type of reflex arc does this represent?
   a. Viscero-somatic
   b. Somato-visceral
   c. Psycho-visceral
   d. Psycho-somatic
   e. Viscero-visceral

2. If there is over-activation of the inferior mesenteric ganglion, what might result?
   a. Increased hepatic glycogenesis
   b. Constipation
   c. Diarrhea
   d. Tachycardia
   e. Oliguria

3. A patient develops hyperactivity of the vagus nerve. What might be an expected outcome?
   a. Miosis
   b. Multiple monthly ovulation events
   c. Priapism
   d. Gastroparesis
   e. Increased salivation

4. A 43-year old patient develops pneumonia. Although the infection is being well controlled and managed with antibiotics, and has not spread to any other locale in the body, what other complaint might he have while he has the pneumonia?
   a. Rhinorrhea
   b. Xerostomia
   c. Anhidrosis of the arm ipsilateral to the infected lung
   d. Cholecystic contractions and spasms
   e. Dermal pain on the lower anterior torso

5. A 62-year old male patient receives surgery of the sigmoid colon to remove an abscess formed secondary to a case of complicated diverticulitis. What other complaints might he have during his convalescence from surgery?
   a. Urinary frequency
   b. Increased seminal vesicle fluid production
   c. Anovulation
   d. Decreased renal function with an increase in BUN and creatinine
   e. Erectile dysfunction

Answers to Chapter 8 Cases

1. E   The patient suffers from chronic rectal spasms. The chronicity of it increases the likelihood for subsequent facilitation and establishment of a reflex arc. If such a reflex is established, the start of it was initiated by pathology of a visceral structure, the rectum. Any body structure has the potential to, through facilitation, cause a "fight or flight" response (sympathetic response) by the body whenever there is damage, injury, or pathology at or in that structure. This is essentially the body's defense system being launched into action. Whatever structure is in a state of pathology has the ability to cause afferent sympathetic stimulation, stimulation that is relayed to the spinal cord. If the stimulation is great or long enough, that segmental level of the spinal cord can become facilitated. Once so, that segment of the cord has the ability to, under little stimulation, cause sympathetic activation to any tissue innervated by that level of the sympathetic nervous system, and/or to stimulate muscles innervated by that same spinal cord segment (to produce muscle hypertonicity and/or spasms), and to cause referred pain to dermatomes and other structures innervated by that segment of the spinal cord. In this case, it is the rectum that is in a state of pathology. The rectum's sympathetic innervation is via T12-L2. Hence, afferent impulses can be carried back to the spinal cord at the segments of T12-L2. If these afferent signals are chronic or of large amplitude, facilitation can be established with a resulting reflex arc. The reflex arc can result in sympathetic stimulation of any structure sympathetically innervated by T12-L2, in hypertonicity or even spasm of any skeletal muscle innervated by T12-L2, and referred pain to any structure innervated by T12-L2, including the lower extremities. The case explains that the patient suffers from excess perspiration of the left lower torso and upper anterior thigh, the areas that are supplied by T12-L2. There is excess perspiration. Perspiration is a sympathetic event. Hence, the sympathetic stimulation to those dermatomal regions is causing excess perspiration. Perspiration is mediated via sweat glands. Sweat glands are glands, namely visceral structures. Thus, this is a viscero-visceral reflex.

2. B   The inferior mesenteric ganglion is the sympathetic collateral ganglion that serves the lower gastrointestinal tract. Over-activation of it will cause sympathetic stimulation to any of the structures of the lower GI tract. In doing this case, one must review all answer options to first identify those structures that are even served in any capacity by the inferior mesenteric ganglion. The liver is served by the celiac ganglion sympathetically, so hepatic metabolism of any sort will not be affected. Constipation and diarrhea are gastrointestinal tract events, so could be considered. Tachycardia results from sympathetic stimulation; the heart is served directly by T1-6 with no intervening ganglion, so it will not be affected in this case. Oliguria would be the sympathetic response of the kidney; but the kidney is supplied sympathetically by T10 and T11, and not by the inferior mesenteric ganglion (which, itself, is supplied by T12-L2). So, the only segment-correct options are constipation or diarrhea. The sympathetic response of the GI tract would be constipation; diarrhea is a parasympathetic response. Remember, parasympathetics are all about normal body function and, if over-activated, "too much" normal function (e.g., diarrhea). Sympathetic responses are always related to whatever will aid the body survive just that particular moment in time, without regard to what encourages survival over a long period of time (which is ruled by the parasympathetic nervous system).

3. B   Hyperactivity of the vagus nerve, also known as cranial nerve X, will cause excess parasympathetic response by any tissues innervated by the vagus. Parasympathetic innervation is responsible for maintaining normal body function and, when in excess, causes "excess" normal body function, so to speak. Thus, in reviewing this case, we must first assess which of the options are served by the vagus. Miosis is caused by pupillary changes; such changes are controlled sympathetically by T1-4 and parasympathetically, as in the case of miosis, by cranial nerve III. There is no vagal contribution. Ovulation occurs in the ovary, and the ovary receives its parasympathetic innervation via the vagus. Priapism is a parasympathetic response by the penis; but the penis receives its parasympathetic innervation from the pelvic splanchnic nerve and not the vagus. Gastroparesis is a gastric event; the stomach is parasympathetically innervated by the vagus. Salivation occurs by way of the salivary glands, structures that receive their parasympathetic innervation via CN VII and IX, and not the vagus. So, of the choices, the only structures parasympathetically innervated by the vagus are the stomach and the ovaries. Gastroparesis, cessation of stomach churning, is a sympathetic response by the stomach. But, ovulation is driven by the parasympathetics (among other things, such as hormones). Thus, excess vagal activity could result in excessive ovulation.

4. B   The lung is sympathetically innervated via T2-4. If there is infection in the lung, it causes pathology in the lung. The pathology can yield facilitation with a reflex arc, resulting in sympathetic stimulation to other structures also innervated by T2-4. In considering the answer options, we must first identify which structures share the same sympathetic innervation. They include the sinus and nasal glands (T1-4), the salivary glands (T1-4), and the arm (non-dermal tissues: T2-8; dermatomes: C4-T2, with sympathetic contribution to C5-C8 possible through cervical sympathetic ganglia (themselves being innervated by thoracic sympathetic elements), and to T1 and T2 via conventional means). The gallbladder is served sympathetically by T6 right, and the dermatome on the lower anterior torso is served by T12. Next, we must consider, of the structure choices provided that are served by T2-4, which ones are demonstrating a sympathetic response. Rhinorrhea is "too much" of a normal body process, and so represents a parasympathetic response by the nasal and sinus mucosa. Xerostomia, or dry mouth, is a sympathetic response by the salivary glands; the normal body process of manufacturing saliva is stopped. Anhidrosis is cessation of sweating; sweating is a sympathetic response, so anhidrosis cannot be the result of the reflex arc herein. Thus, the patient could develop xerostomia.

5. E   The abscess alone would be capable of establishing facilitation, but surgery, while used as a treatment, causes extreme sympathetic responses due to the invasive nature of the process. Thus, the surgery and the following convalescent period could be accompanied by facilitation involving the segments that normally serve the sigmoid colon sympathetically, namely T12-L2. Again, we must assess the answer options to determine what structures among the choices given are served sympathetically by T12-L2. The urinary bladder, the seminal vesicle, and the penis fall into that category. On the other hand, the ovary is supplied sympathetically by T10-11; the ovary would be inconsistent in this case anyway since this patient is male. The kidneys are served sympathetically by T10-11, too. Of the structures for which a reflex arc of sympathetic stimulation could have been established, one must determine which of the choices reflect a genuine sympathetic response. Urinary frequency and increased seminal vesicle fluid production are both parasympathetic responses, whereas erectile dysfunction is a sympathetic response by the penis. This man can be expected to develop a temporary case of erectile dysfunction while convalescing from his sigmoid colon surgery.

Chapter Nine:
VISCERAL AND SYSTEMIC CONSIDERATIONS

Disease and dysfunction can occur at any place within the human body. Such dysfunction and disease alters the local, regional, and, sometimes, systemic environment within and of the body. We already are aware that visceral dysfunction can cause somatic symptomology through viscero-somatic reflexes, and that somatic dysfunction can cause visceral symptomology and dysfunction simply through somato-visceral reflexes. In fact, constant stimulation of the spinal cord by visceral or somatic afferents facilitates those particular spinal cord segments, thereby reducing the threshold in those spinal cord segments. Such segments are said to be facilitated. With a lowered threshold, little excitation is required to stimulate any structure innervated by that facilitated segment. As a result, assessable and often symptomatic somatic or visceral dysfunction can result secondary to the primary insult. Unfortunately, once such a facilitated segment is established, continued excitation to that segment or any kind of additional stress, be it physical, emotional, or mental, can easily and rapidly incur an explosion of sympathetic impulses to the facilitated segment's associated viscera and/or somatic structures.

Facilitation is common with chronic dysfunction and infrequent with acute dysfunction.

Treatment of any type of systemic dysfunction has seven treatment objectives within the osteopathic manipulative treatment plan, as displayed below in order of execution:

1. Support homeostatic mechanisms.

2. Address issues and provide treatment where necessary to ensure that the body has a healthy metabolism and can properly use the nutrients provided to it.

3. Support the entire patient, including mentally and emotionally.

4. **Treat the dysfunction responsible for initiating the facilitated spinal cord segment!!**

5. **Reduce the sympathetic nervous system contribution to the problem:**
    a. utilize Chapman's reflexes
    b. treat the sympathetic chain ganglia
    c. treat collateral ganglia (preaortic and cervical ganglia)

6. **Aid in the movement of fluids, particularly the movement of lymphatic fluids back to the intravascular space as well as interstitial fluids back to the circulation:**
   - a. release thoracic inlets, releasing the cervicothoracic diaphragm
   - b. release the abdominal diaphragm
   - c. treat the mesenteries
   - d. release the pelvic diaphragm
   - e. utilize lymphatic pumps
   - f. release the craniocervical diaphragm
   - g. treat fascial restriction

7. **Effect a parasympathetic nervous system balance:**
   - a. treat the OA (and, possibly, the AA and C2)
   - b. reduce any sacral dysfunction
   - c. treat the pelvic splanchnic nerves
   - d. treat cranial nerves III, VII, IX, X
   - e. address any somatic dysfunction associated with the sphenopalatine ganglion
   - f. treat the cranium

In fact, the actual manipulation usually **first addresses and treats the cause for the facilitation**. Manipulation is also used to **reduce sympathetic contribution**. Then, the **lymphatics are addressed**. And, finally, treatment geared towards attaining an **improved parasympathetic balance** is performed. In most cases, this means that the primary dysfunction is treated along with treatment directed towards T1 through 12 and the ribs. Then, lymphatic pumps and associated treatments may be utilized. And, lastly, treatment is usually completed with manipulation often focused on the OA and sacrum.

In order to be able to fully understand the treatments for the various groups of organs, one must be fully familiar with the neural connections as described in the chapter, "Innervation and Neural Connections." A more detailed organ-specific outline is provided at the end of this chapter, along with a body diagram of the innervations. It is essential that one be expertly versed in these neural connections. However, despite these recognized neural connections and their associated reflexes, be aware that **any very intense visceral or somatic afferent activity may actually result in a spread of facilitated segments cephalad and/or caudad to the expected thoracolumbar segment. This may result in palpable (and possibly symptomatic) somatic dysfunction of tissues associated with those more caudad or cephalad segments**. These dysfunctions may represent viscero-visceral, viscero-somatic, somato-visceral, and/or somato-somatic reflexes. In extreme cases, cephalad or caudad facilitation may also be seen in psycho-visceral or psychosomatic reflexes. Hence, it is critical to understand and know the direct neural connections and reflexes and to appreciate the fact that any of these may result in the creation of a facilitated segment. Furthermore, once a segment becomes facilitated and if the afferent impulses are intense or prolonged enough, more facilitated segments may form caudad or cephalad to the original facilitated segment.

So, first treat the cause(s) for the facilitation.

Then, quell the sympathetic contribution. The primary manipulative method to moderate hypersympathetic activity is **rib raising.** Rib raising also provides the added benefit of improving diaphragmatic action which, of course, aids the lymphatics. The technique functions to lift and rotate the heads of the ribs, thus causing tension on the fascia that is associated with both that rib head and the nearby sympathetic chain ganglion. This initially stimulates the ganglia, causing even more sympathetic output. But, this effect is short-lived because an inhibitory process, modulated by the medulla, is set into place, resulting in inhibition of the sympathetic system.

In the most basic terms, rib raising is performed on the supine patient while the practitioner is sitting to the side of the patient. The practitioner's hands are inserted underneath the patient, making contact with the angles of the ribs with the finger pads of the fingers. The practitioner lifts up (towards ceiling), moving the ribs (one should also see the sternum move). Then, a gentle lateral traction is applied. Throughout the process, the dorsum of the MCP joints is used as a fulcrum and the forearms are well planted against the surface of the table.

If one wanted to have the effects of rib raising in the lumbar area (where there are sympathetic chain ganglia, but no ribs!), one may perform the **"Ileus Prevention Treatment"** (also known as the **"Inhibition Technique for Lumbar Segments").** In this method, the clinician again sits beside the supine patient and places both of his hands under the paraspinal muscles in the affected lumbar area (on the same side as the clinician). The finger pads should be placed just medial to the paraspinal muscle mass and the "heel" of the hand (hypothenar and thenar areas) should be lateral to that muscle mass. Then the clinician lifts his fingers (towards the ceiling) and brings them laterally, doing so strongly enough to cause a lumbar extension. This position is maintained for 60-90 seconds.

After rib raising or the inhibition technique for lumbar segments is performed, further sympathetic decreases may be desired. In this case, the **collateral ganglia** may be assessed and, if needed, treated. These ganglia include the preaortic and cervical ganglia. The **preaortic ganglia** are the **celiac, the superior mesenteric and the inferior mesenteric ganglia,** and are all located just anterior to the aorta in the abdomen. Notice, as in general OMT, that the central structures (i.e., the sympathetic chain) are treated before the peripheral structures (i.e., the ganglia). The celiac ganglion innervates the upper GI tract (stomach, duodenum, liver, gallbladder, spleen, and pancreas). The superior mesenteric ganglion innervates the small bowel distal to the duodenum, the ascending colon, the proximal transverse colon, the kidneys and adrenal glands, as well as the gonads. The inferior mesenteric ganglion innervates the distal transverse colon, the descending colon, the sigmoid colon, and the rectum, as well as the pelvic organs (excluding the gonads).

The preaortic ganglia are first assessed by having the clinician place both hands on the abdomen, with fingers extended. The longer fingers are slightly flexed to ensure that the finger pads of all fingers form a straight line along the patient's abdominal midline. Then, pressure is applied through the finger pads. This is done over the area of each ganglion. Facilitated ganglia will produce a rapid objective sensation to the clinician of resistance, while the patient will complain of tenderness with even mild or moderate pressure. There may also be symptomology associated with the organs innervated by that ganglion.

If these ganglia prove to be facilitated, the **"Ganglion Inhibition Treatment"** may be undertaken. In this treatment, the hands and fingers are positioned exactly as they were for the ganglion assessment. The patient takes a deep breath and, with exhalation, the clinician follows the excursion, pressing deeply into the tissues until resistance is met. This position is maintained. The breathing cycle is repeated again with the further following of excursion even deeper by the clinician. The pressure is held until resistance is relaxed or for about 90 seconds.

Besides the preaortic ganglia, the other collateral ganglia are the **cervical ganglia.** They are particularly important to the heart and structures of the head (particularly the sinuses) and the neck. Anatomically, they are comprised of 2 to 3 ganglia on each side of the neck (the superior, middle, and, sometimes, inferior cervical ganglia). They represent the original 8 cervical paraspinal ganglia that coalesced together.

The superior cervical ganglion lies anterior to the transverse processes of C2, 3, and 4, and supplies sympathetics to C1-4 nerves as well as to the cardiac plexus. The middle cervical ganglion lies at the C6 vertebra. It provides sympathetics for cervical nerves C5 and C6, as well as to the cardiac plexus. The inferior cervical ganglion is located on the anterior surface of the head of the first rib. It provides sympathetic innervation to C7, C8, and T1, and, again, to the cardiac plexus. (Can you see why the cervical ganglia are so important to the heart?!)

The fascia of the cervical region is closely associated with these ganglia as well as the cervical joints. So intimately involved are the cervical joints to these ganglia that any cervical somatic dysfunction of these joints or the associated fascial structures can result in innumerable visceral problems secondary to cervical ganglia stimulation. These problems may include sinusitis, eye dysfunction, ear dysfunction, cardiac tachyarrhythmias, and other cardiac manifestations. Many times, these problems and their associated cervical ganglion hyperstimulation counterparts can be corrected by freeing cervical fascial planes and treating any cervical joint somatic dysfunction.

Chapman's reflexes, developed by Frank Chapman, DO, may also be used to moderate sympathetic hyperstimulation. They are well-defined points of pain that result from viscero-somatic reflexes. Thus, Chapman's points are used to identify tender points on the body that result from a facilitation of a segment; and, that facilitation, in this case, has frequently occurred secondary to particular visceral dysfunction. There are both anterior and posterior points. So, tenderness felt upon mild to moderate pressure at any of these Chapman's points is often related to dysfunction of a specific organ. However, always remember, like any diagnostic tool, all results must be correlated with other historical and physical findings in the patient in order to ensure the most accurate diagnosis. In fact, if a positive Chapman's point is found, the next step is not treatment of that point, but rather further history and physical examination as related to the organ possibly involved. Although rib raising is often the first technique used to modulate sympathetic input, one should be aware that a good <u>Chapman's assessment is actually performed **BEFORE** any manipulation has been done that would alter sympathetic pathways</u>.

Therefore, if one anticipates using Chapman's reflexes, the Chapman's assessment must be the first step taken in addressing reflex responses.

The Anterior Chapman's Points, as they are more sensitive than the posterior points, **are used for diagnostic purposes. The Posterior Chapman's Points are used for treatment**. Thus, assessment is done via Anterior Chapman's Points prior to the performance of any manipulation, especially sympatheticomodulatory techniques. Then, any treatments (including rib raising) may be performed. At some point, the Chapman's reflexes may be treated; typically they are treated after rib raising. However, BEFORE the Chapman's reflexes are treated, all pelvic somatic dysfunctions must be treated. After that, the Posterior Chapman's Points are treated by pressing firmly over the soft tissues at the tender point, engaging a circular motion for 20 to 30 seconds. In practice, the use of Chapman's points for treatment is not as commonly employed as other medical treatments for the organs involved.

Clinical Note: The exception to the rule of "not treating Anterior Chapman's Points" is the anterior tibial band Chapman's points. These may be treated directly to normalize the colon in such disorders as irritable bowel syndrome, ulcerative colitis, and viral diarrhea.

It should be noted that Chapman's reflexes are distinctly different from Travell's myofascial trigger points and Jones Tender Points. Travell's myofascial trigger points, described by Janet Travell, MD, are focal areas of irritation on the body that are usually associated with hypertonic skeletal muscle or taut fascia; when compressed, pain or autonomic response elsewhere (referred pain or referred neural response) is generated – sometimes this referred pain or autonomic response is generated spontaneously by these points without the contribution of compression. They arise from somato-somatic, somato-visceral, or, occasionally, viscero-somatic reflex. More common treatments used for trigger points that are somatically generated are vapocoolant spray with stretch, focal anesthetic injection, muscle energy, or myofascial release.

Jones Tenderpoints, described by Lawrence Jones, DO, are small areas of focal irritability in muscles and fascia that are painful on compression. Like trigger points, they are usually associated with hypertonic muscle and/or taut fascia. But, unlike trigger points, they do not refer pain or refer an autonomic response. Their primary purpose is in the evaluation of myofascial structures for the identification of focal myofascial dysfunction, and as a tool in monitoring myofascial pain improvement with Counterstrain technique. Likewise, tenderpoints are treated with Counterstrain technique.

Site of Reflex Point (areas of pain)	Dysfunctional or Pathologic Structure
Medial inferior clavicle	Sinuses
Medial superior clavicle	Middle ear
Superior lateral edge of manubrium	Pharynx
Middle lateral edge of manubrium	Tonsils
2nd costosternal joint	Tongue
Superior aspect of 3rd costosternal joint	Esophagus, heart
5th costochondral joint, LEFT	Stomach (parietal)
5th costochondral joint, RIGHT	Liver
6th costochondral joint, LEFT	Stomach (peristalsis)
6th costochondral joint, RIGHT	Liver, gallbladder
7th costochondral joint, LEFT	Spleen
7th costochondral joint, RIGHT	Pancreas
8th, 9th, and 10th costochondral joints	Small intestine
Tip of 12th rib, RIGHT	Appendix
Superior edge of pubic bone just lateral to pubic symphysis	Ovary
Anterior thigh from area proximal to knee to area just distal to greater trochanter	Colon
Lateral thigh from area proximal to knee to area just distal to greater trochanter	Prostate
Lesser trochanter of femur	Rectum
Lateral superior humerus	Retina, conjunctiva
Medial superior humerus	Neck
Tip of the coracoid process	Cerebellum

After the sympathetics are treated in some manner to reduce hypersympathetic involvement, the lymphatics should be addressed. Remember, they were already addressed in some measure by the rib raising, which also serves to help in good diaphragmatic action (and, therefore, good lymphatic pumping).

The **cervicothoracic diaphragm** (Sibson's fascia) is the common pathway in lymph drainage from anywhere in the body (the right lymphatic duct and the thoracic duct run through it - the thoracic duct twice and the right lymphatic duct only once). This diaphragm should, therefore, always be evaluated. It is the most common site for obstructed lymphatic flow. It is the fascial diaphragm at the thoracic inlet. Any fascial preference or somatic dysfunction in this area should be treated, and the inlet should be freed.

The anatomic thoracic inlet is framed by the manubrium, the first rib, and the first thoracic vertebra, and the apex of the lung protrudes through it. The **"functional thoracic inlet"** is the **"clinical thoracic inlet,"** and is defined as being framed by the manubrium with the angle of Louis, the first two ribs on each side, and the first four thoracic vertebrae. The latter definition is what is most commonly used in osteopathy.

After the **cervicothoracic diaphragm** (the most important one) is assessed and treated, the other three diaphragms should be assessed and treated as necessary:

Abdominal Diaphragm
Pelvic Diaphragm
Craniocervical Diaphragm

Cranial OMT is used to free the craniocervical diaphragm.

Abdominal diaphragm dysfunction is manifested clinically by the lack of movement of any of the abdominal tissues down to the pubic symphysis during respiratory effort. Further evaluation of the abdominal diaphragm can be performed by placing a flat hand over the epigastric area and the other hand posteriorly at the thoracolumbar junction, and pressing and palpating for preference of rotation about an AP axis and restriction associated with it. To treat the abdominal diaphragm, the thoracolumbar paraspinal musculature is relaxed by soft tissue techniques, stretching, and myofascial release. Then the diaphragm is domed. Lasting effects of a redoming will only occur if somatic dysfunction of the diaphragmatic attachments has been treated: L1-3 (including ANY lordosis), ribs 6-12, and the xiphoid process. It is prudent to also assess and treat the origin of the innervation for the diaphragm (C3, 4, 5). Any quadratus lumborum and iliopsoas dysfunction may also interfere with appropriate diaphragmatic motion and response to treatment.

The pelvic diaphragm is diagnosed by pressing into the lateral sides of the ischiorectal fossa, or by evaluating the muscles of the pelvic diaphragm via a rectal or vaginal examination. The pelvic diaphragm can then be released through the perineum, the rectum, or the vagina. All pelvic diaphragm manipulative treatment should be followed by at-home Kegel exercises by the patient.

Specifically, to treat the pelvic area, the ischiorectal fossa is released by having the patient lie supine with hips and knees flexed to 90 degrees. The clinician places the fingers of one hand into the ischiorectal fossa fat, and places the other hand on the patient's hip for counterforce. Then, the fingertips are pressed superiorly and laterally against the margin of the fossa. The patient inhales and exhales; with each breathing cycle of exhalation, the fingers drift more superiorly following the diaphragm. The counterforce of the pelvic diaphragm is resisted during the patient's inhalation.

After all four diaphragms are assessed and treated/freed as necessary, lymphatic pumps should be employed to encourage lymphatic flow and eventual lymphatic return to the circulation. Some lymphatic pump techniques include:

Thoracic Lymphatic Pump Treatment
Pedal Lymphatic Pump (Dalyrimple)
Pectoral Traction
Splenic Pump
Liver Pump
Posterior Axillary Fold Technique
Lower Extremity Fascial Pathway Treatment

The **thoracic lymphatic pump treatment** engages a rapid, strong negative intrathoracic pressure that encourages rapid lymphatic return to the thorax from other parts of the body. The patient lies supine while the clinician places steady pressure via the hands onto the patient's chest (hands are placed on upper ribs in mid-clavicular area). Steady downward pressure is applied during the patient's exhalation. A vibratory motion may also be applied during exhalation to encourage movement of respiratory secretions. Thoracic movement is discouraged by maintaining the hand pressure during the patient's inhalation phase. The pressure is again increased as the patient starts another exhalation. Then, just shortly after inhalation is initiated, the clinician SUDDENLY releases the pressure, causing a rapid inhalation of air as well as influx of lymph.

With the **pedal pump,** the patient is supine and the clinician intermittently and rhythmically applies force through the feet that ultimately causes abdominal movement. The rhythm usually occurs at a rate of approximately 90-120 forces per minute. This method may be performed on the dorsiflexed or plantar flexed foot. Dorsiflexion is believed to most aid the lumbar and lumbosacral regions, while plantar flexion is associated with more aid to the thoracic and cervical regions, as well as the rib area.

The **splenic pump** is used to move lymph in patients with systemic infections and/or anemia. With the patient supine, gentle alternate compression and release is applied to the area over the spleen. This must not be done during splenomegaly.

The **liver pump** is performed in the same manner as the splenic pump, except that it is performed over the area of the liver. It decongests the lymphatic and venous systems of the liver, relieving visceral congestion and, theoretically, aiding in the overall detoxification process.

After the lymphatic pumps have been performed, rib raising may be performed again to encourage abdominal diaphragmatic excursion and to, once again, decrease hypersympathetic tone. The hypersympathetic tone constricts veins and larger lymphatic vessels. Such constriction prevents proper lymphatic drainage and encourages tissue congestion.

Fascial dysfunction, beyond that which may be associated with the diaphragms, should also be assessed and treated so as to encourage good lymphatic flow. Fascias should be examined for evidence of compromised homeostatic mechanisms. **Compensated patterns,** as described by G. Zink, DO, alternate with each other, whereas **uncompensated patterns** consist of all fascia running in the same direction.

To diagnose fascial uncompensation, one must palpate the anatomic transition sites of the fascia:

occipitoatlantal area
cervicothoracic area
thoracolumbar area
lumbosacral area

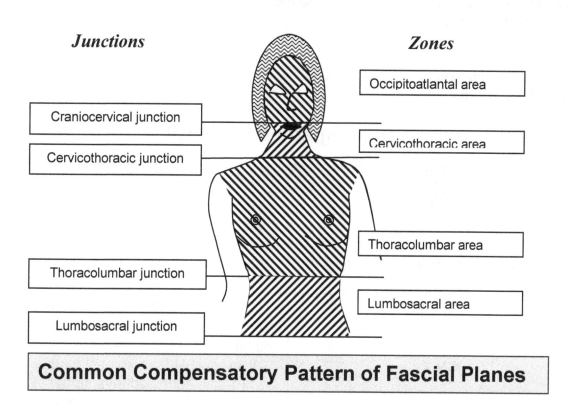

Junctions   *Zones*

Craniocervical junction

Cervicothoracic junction

Occipitoatlantal area

Cervicothoracic area

Thoracolumbar area

Thoracolumbar junction

Lumbosacral area

Lumbosacral junction

Common Compensatory Pattern of Fascial Planes

Alternation of fascial patterns should be able to be palpated; these alternating patterns are compensatory fascial patterns, as demonstrated above. In addition to this alternating pattern, 80% of people have a fascial rotation to the left at the OA area (so the face is designated as having R_L pattern, representing fascial torsion from right to left), to the right at the cervicothoracic area (designated as an L_R pattern representing fascial rotation or torsion from the left to the right), to the left at the thoracolumbar area (designated as R_L), and to the right at the lumbosacral area (designated as L_R). This compensatory pattern, as described herein and as depicted in the previous illustration can be said to be left, right, left, right, or "L,R,L,R". Such a pattern is termed the **"Common Compensatory Pattern (CCP)."** Persons who have a compensatory pattern but one that is just the opposite of the aforementioned pattern are said to have the **"Uncommon Compensatory Pattern."** Most non-compensated (i.e., non-alternating) fascial patterns are of traumatic origin. The originating dysfunction must be treated.

To aid in lymphatic flow, the **mesenteries** may also be treated. In basic terms, the mesenteries are placed at right angles to their attachments, and gentle tension is applied. The patient is then instructed to take a shallow breath and hold it as long as possible. The mesentery should be a bit loosened by the next breath. This is repeated 2-3 times.

The <u>final step</u> in treating systemic dysfunction is to treat **parasympathetic nervous system dysfunction.** The parasympathetics control viscera by way of cranial nerves III, VII, IX, and X, as well as through the pelvic splanchnic nerves from S2, 3, and 4. The parasympathetic nerves synapse in various ganglia and plexuses: ciliary, sphenopalatine, otic, submandibular, submaxillary, and myenteric (Auerbach's) and submucosal (Meissner's) plexuses of the GI tract. Also, parasympathetic fibers in the form of the pelvic splanchnics along with preganglionic sympathetic fibers from T12-L2 contribute to the hypogastric plexus; this is a plexus located in the lumbosacral area.

The following table depicts the tissues affected by the aforementioned parasympathetic nervous system (PNS) structures:

Parasympathetic Control

Conduit for PNS Control	Affected Tissue
CN III	Eye
CN VII (CN VII synapses on the sphenopalatine ganglion)	Glands on the mucous membranes of the sinuses and nasal mucosa, as well as the pharynx, Eustachian tubes, and lacrimal glands
CN IX and X	Carotid body and carotid sinus
CN X	GI tract from inferior 2/3 of esophagus to distal 1/3 of transverse colon, pancreas, liver, gallbladder, spleen, lungs and bronchi, heart, kidneys, upper ½ of ureters, testes, ovaries
S2, 3, 4 (pelvic splanchnic nerve)	Left colon, sigmoid colon, rectum, all pelvic organs except gonads

Treating the **sphenopalatine ganglion** can have a strong influence on modulating glandular secretions in the head. It is also known as the pterygopalatine ganglion, and is the ganglion that is a conduit for parasympathetic impulses via CN VII. However, the ganglion also houses sympathetic efferent (postganglionic) fibers that are provided to the ganglion via the nerve of the pterygoid canal and that carry impulses from thoracic sympathetic elements. Thus, modulation of this ganglion can have tremendous balancing effects with respect to glandular secretions in the head. Unfortunately, modulation of this ganglion provides somewhat of a challenge since the ganglion cannot be palpated by the finger. It is located in the sphenopalatine fossa of the skull, and is inferior to the maxillary division of CN V. Nonetheless, it may be influenced by manipulation of the pterygopalatine muscle fascias. These may be manipulated through the open mouth of the patient to modulate glandular secretions of the head. The clinician passes his/her finger over the molars of the upper jaw, and upon reaching the maxillary ridge posteriorly, moves his finger to the lateral edge of the maxillary ridge. Then the finger is moved cephalad over the pterygoid plates and pressure is applied. Then the patient is asked to nod forward toward the palpating finger. This is repeatedly 2-3 times on each side of the mouth.

Treatment of the **vagus nerve** is the KEY TOOL in balancing parasympathetic stimulation of the thoracic and abdominal viscera (including the GI tract proximal to the left colon). Manipulation of the OA is particularly useful; success may also be had by manipulation of the AA and C2 joints.

The **pelvic splanchnic nerves** may also be treated to modulate parasympathetic influence of the left colon, sigmoid, and rectum, and the pelvic organs. This is accomplished through indirect inhibition of the inferior mesenteric ganglion and balancing pelvic splanchnic output by way of hypogastric plexus modulation. Remember, the hypogastric plexus contains sympathetic fibers from T12-L2 and parasympathetic fibers from the pelvis splanchnic nerves, and lies anterior to the lumbosacral region. As such, sacral rocking is able to modulate the fibers in this plexus. The patient is placed in the prone position, and the clinician places one hand over the sacrum. The other hand is placed over the first hand. The sacrum is then rocked with respiration. This is termed **"Rocking the Sacrum."** A similar technique can be done with the patient supine, and one hand of the clinician under the patient at the location of the sacrum. If this is to have lasting effect (if any effect at all), all somatic dysfunction of the sacrum and innominates must be assessed and treated, too.

In all cases, always treat the patient, not the dysfunction. Always ensure that the treatment plan is individualized to each patient's needs and problems. And, be aware that, no matter the techniques one uses, in these cases one is dealing with a primary stressor, a facilitated segment, and a secondary dysfunction - all components of the reflex arc.

DETAILED LIST OF
SEGMENTAL SYMPATHETIC INNERVATION

Tissue	Segments	Tissue	Segments
Sinuses	T1-4	Small intestine	T10-11
Eustachian tube	T1-4	Colon	T10-L2
Lacrimal glands	T1-4	Right Colon	T10-11
Salivary glands	T1-4	Appendix	T12
Thyroid	T1-4	Left colon	T12-L2
Trachea	T1-6	Adrenal glands	T10-11
Bronchi	T1-6	Kidney	T10-11
Lower 2/3 of esophagus	T5-6	Upper ureter	T10-11
Aortic arch	T1-5	Lower ureter	T12-L1
Heart	T1-6	Bladder body	T12-L2
Lungs	T2-4	Trigone/sphincter	T12-L2
Lung visceral pleura	T2-4	Ovary and testis	T10-11
Stomach	T5-9 left	Uterus	T12-L2
Duodenum	T5-9	Prostate	T12-L2
Liver	T5 right	Genital cavernous tissue	L2
Gallbladder	T6 right	Mammary glands	T1-6
Biliary tree	T6 right	Arms	T2-8
Spleen	T7 left	Legs	T11-L2
Pancreas	T7 right	Penis, seminal vesicle	L2

Be aware that there is much variation in sympathetic innervation reported in the literature. Every attempt has been made to list those innervations most accepted in the field and most widely relied upon in standardized national testing. Rather than learning all of the variations published in the literature, it is recommended that you learn one set of reported innervations thoroughly and use that as the basis for your work and academic examinations. Also, one must recognize that any reflexes occurring by way of facilitation can, via spread in the spinal cord, result in stimulation from segments cephalad or caudad to the originally facilitated segment.

SYMPATHETIC MAN
An Illustration of Regional Sympathetic Innervation

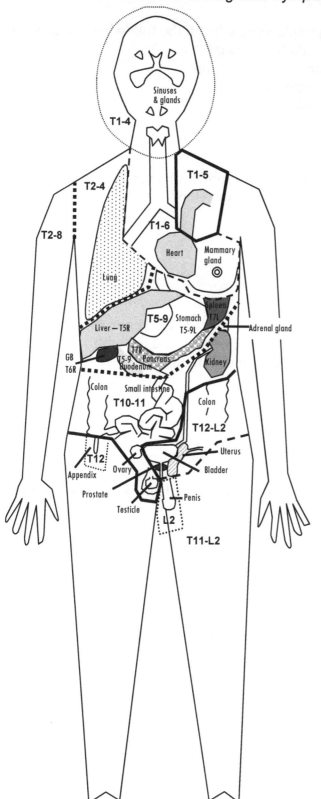

Key:

GB = gallbladder
L = left; R= right

Face, sinuses, lacrimal and salivary glands, pharynx, larynx: T1-4

Trachea, bronchi, heart: T1-6

Aortic arch: T1-5

Lung: T2-4

Liver: T5R

Mammary gland: T1-6

Stomach: T5-9L

Duodenum: T5-9

Gallbladder: T6R

Biliary tree: T6R

Pancreas: T7R

Spleen: T7L

Small intestine (except duodenum), right and transverse colon, kidney, upper ureter, adrenal gland: T10-11

Appendix: T12

Ovary and Testis: T10-11

Lower ureter, bladder, urethra, prostate gland, uterus, fallopian tubes, descending and sigmoid colon, rectum: T12-L2

Penis, seminal vesicle: L2

Upper extremity: T2-8

Lower extremity: T11-L2

Chapter 9 Review Cases

1. A 23-year old patient develops splenomegaly. Of the following, which viscero-visceral reflex would be most likely to occur?
   a. Decreased gastric secretions and contractions
   b. Pain along the left T7 dermatome
   c. Increased pancreatic secretions
   d. Constipation
   e. Tachycardia

2. A 68-year old female patient develops a chronic infection in the umbilicus. As a result, what other ailments might she also develop?
   a. Polyuria with hypernatremia
   b. Anovulation
   c. Urinary retention
   d. Diarrhea
   e. Hyperkalemia

3. A 33-year old male suffers a motor vehicle accident wherein the right quadriceps are severely damaged, requiring weeks of healing and rehabilitation. What other dysfunction would be most likely to result from this?
   a. Anhidrosis of the right leg
   b. Erectile dysfunction
   c. Urinary frequency
   d. Intestinal hypermotility
   e. Azoospermia

4. A patient develops hyperactivity of the parasympathetic nervous system via the vagus. What may result?
   a. Spontaneous ejaculation
   b. Fecal incontinence
   c. Hyperovulation
   d. Increased hepatic gluconeogenesis and glycogenolysis
   e. Hypersalivation

5. You treat an anterior Chapman's point found at the medial inferior clavicle. What did this likely resolve?
   a. Hypersalivation
   b. Flatulence
   c. Galactorrhea
   d. Sinus discomfort
   e. Gonorrhea

Answers to Chapter 9 Cases

1. A   Splenomegaly is an enlarged spleen, and represents pathology of the spleen. Any injury or pathology of tissue can activate the body's defense mechanisms, namely the sympathetic nervous system. The result can be facilitation and the development of reflex arcs. The sympathetic innervation to the spleen is via T7 left. Hence, splenomegaly could result in sympathetic stimulation to other tissues sympathetically innervated by T7 left. In reviewing the answer choice options, the only structures listed that are served by T7 are the stomach and the T7 dermatome. The stomach problem noted is decreased gastric secretions and contractions, which would be a gastric response to sympathetic stimulation. On the other hand, pain in the T7 dermatome could be referred pain also generated from facilitation of the T7 segment. Thus, we have to consider the question in terms of which of these represents a viscero-visceral reflex. Since it is generated by the spleen, it originated with a visceral element. However, referred pain, as it is due to activation of pain fibers, does not constitute a visceral response. It is a somatic response since nervous system structures such as nerves are considered to be somatic structures. Pain is conducted by nerves. On the other hand, the stomach is a visceral structure, and so the decreased gastric secretions and contractions in response to the splenomegaly would be consistent with a viscero-visceral reflex.

2. E   A chronic infection of any sort is a prime candidate for initiating a reflex arc through facilitation. This infection is occurring in the T10 dermatome. Thus, structures sympathetically innervated by T10 can receive inappropriate sympathetic stimulation. The T10 innervated structures, of the ones provided in the answer options, include kidney, ovary, bowel, and adrenal gland. Polyuria could be due to the kidney or the adrenal gland. If due to the kidney, it would be a parasympathetic and not a sympathetic response. On the other hand, sympathetic stimulation of the adrenal gland CORTEX results in decreased manufacture and release of cortical hormones, including the mineralocorticoid known as aldosterone. Do not confuse this with the well-established increase in cortisol, another cortical hormone, that occurs during systemic stress; such systemic stress causes release of high quantities of ACTH which, in turn, stimulate the adrenal gland to manufacture and release increased amounts of cortisol. However, direct sympathetic stimulation to the adrenal gland alone, with no anomalous ACTH influence due to systemic stress, will result in decreased cortical hormone production, including decreased aldosterone. The decreased aldosterone causes decreased Na-K exchange at the level of the distal convoluted tubule and collecting duct, resulting in polyuria and hypOnatremia. Thus, polyuria with hypernatremia is neither a sympathetic response by the kidney or the adrenal gland. Anovulation is a sympathetic response by the ovary to sympathetic stimulation. However, it is irrelevant herein since the woman is well beyond her reproductive years and has long since, undoubtedly, already stopped ovulating. While the bowel is also, in part, sympathetically innervated by T10, sympathetic stimulation would cause constipation, not diarrhea. The last consideration is hyperkalemia. Sympathetic stimulation to the adrenal gland, as already established, will cause a decrease in aldosterone production, which will cause a decrease in Na-K exchange in the kidney. This will result in polyuria, hyponatremia, and retention of potassium known as hyperkalemia. Thus, the patient might develop hyperkalemia.

3. B   Damage to any tissues can launch the body's defense mechanisms, including the sympathetic nervous system. The result can be facilitation with reflex arcs to tissues sharing the same innervation. The quadriceps are innervated by L2, L3, and L4, while overall sympathetic innervation to the leg as a whole is provided by T11-L2. Thus, damage to the quadriceps could easily activate the sympathetics from T11-L2, with greatest effect on L2 (sympathetic contribution and involvement at the spinal cord level ends at L2). In considering the answer options, the leg, the penis, the bladder, and the intestine are all innervated by T12-L2. Azoospermia, the lack of sperm production, is a sympathetic response by the testis, but the testis is sympathetically innervated by T10-11. Of the structures that receive their innervation from L2, the leg can be affected, but not by way of anhidrosis. Anhidrosis is a parasympathetic response. The penis responds sympathetically with ejaculation; but sympathetics that override parasympathetics result in the inability to have an erection, causing erectile dysfunction. The urinary bladder responds to sympathetic stimulation by experiencing urinary retention, not frequency. Intestinal hypermotility is not a sympathetic response of the intestine as that is more consistent with parasympathetic stimulation. Thus, of the choices provided, the only possible one is erectile dysfunction.

4. C   The only structures, of the ones noted, that are innervated by the vagus are the ovary and the liver, since the rectum and penis are parasympathetically innervated by the pelvic splanchnic nerve and the salivary glands are parasympathetically controlled by CN VII and IX. However, vagal (parasympathetic) stimulation of the liver encourages glycogenesis; sympathetic stimulation results in increased gluconeogenesis and glycogenolysis to provide glucose to tissues during times of pathology or injury. Vagal stimulation to the ovary causes ovulation. Thus, hyperovulation could be a result.

5. E   That Chapman's point is reflective of pathology in the sinuses. Thus, treatment of such may resolve any sinus discomfort that exists.

Chapter Ten: CRANIOSACRAL PRINCIPLES

William Gainer Sutherland, DO, DSci (Hon) developed the cranial field. His theories state that the central nervous system, the cerebrospinal fluid, and the dural membranes function as a single unit. This single unit is termed the **primary respiratory mechanism (PRM)**. Sutherland further theorized that the PRM actually controls and regulates the pulmonary respiration, known in the cranial field as the "secondary respiration."

By definition, as espoused by Sutherland, there are five anatomic and physiologic components of the PRM:

1. Mobility of the cranial and spinal membranes
2. Fluctuant CSF
3. Motility of the CNS (brain and spinal cord)
4. Mobility of the cranial bone articulations
5. Involuntary mobility of the sacrum

The dura mater is the outermost membrane of the CNS. It surrounds the entire CNS and has firm attachments to the cranial bones as well as to the foramen magnum, C2, C3, and S2. Thus, all inherent motion of the CNS or fluctuation of the CSF theoretically causes these membranes to move. Accordingly, such movement is believed to cause cranial bones to move in response to such CNS or CSF movement. Sutherland termed this dural membrane the **Reciprocal Tension Membrane (RTM)**.

Remember, the RTM is attached to several areas, including to S2. In fact, it is specifically attached to the posterior superior aspect of the spinal canal within S2. Thus, according to the cranial field philosophy, any RTM movement will also cause movement of the sacrum. As such, this movement occurs about a transverse axis that exists through S2. This axis is known as the respiratory axis.

The key articulation in the cranial field is the **sphenobasilar synchondrosis (SBS)**. It moves through flexion and extension in response to pull by the RTM.

There are 6 types of SBS strains:

1. Flexion and extension
2. Torsion
3. Sidebending/rotation
4. Vertical strain
5. Lateral strain
6. Compression

Of these, flexion, extension, torsions, and sidebending/rotations are all considered physiologic unless their presence causes dysfunction or limits normal flexion/extension movement associated with the **CRI (cranial rhythm impulse)**. The latter are rhythmic impulses that Sutherland describes in his theories; he described these impulses as reflecting the pulsation of the CSF. These are supposed to be able to be palpated when palpating the skull. Normal CRI is considered 10-14 cycles/minute.

FLEXION - during flexion, all unpaired cranial bones ("midline bones") move into flEXion, and all paired cranial bones move into EXternal rotation. Flexion causes the RTM to be pulled cephalad, thus tugging on the sacrum at S2 to incur backward bending of the sacral base. Such backward bending in the cranial field is known as counternutation. Flexion causes the head to change shape, too: WIDE HEAD, DECREASED AP DIAMETER

EXTENSION - When the SBS moves into extension, the paired cranial bones internally rotate. Furthermore, the taughtness on the RTM is released, allowing the sacrum to dip forward into nutation. Extension causes the head to change shape: NARROW HEAD, INCREASED AP DIAMETER.

Midsagittal Section of Skull

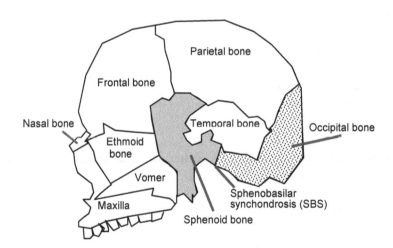

Craniosacral Flexion and Extension Dysfunction

Depicted below is craniosacral flexion and extension (lateral view); for illustrative purposes and to allow for better understanding of the positions undertaken in these dysfunctions, the flexion and extension is shown in exaggerated examples.

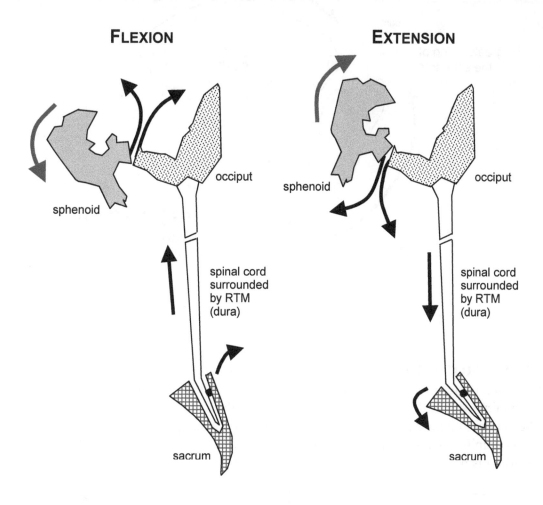

TORSION - a twisting of the SBS occurs whereby the sphenoid and other anterior cranial structures rotate in one direction while the occiput and other posterior cranial structures rotate in the opposite direction. The type of torsion is determined by which greater wing of the sphenoid is more superior.

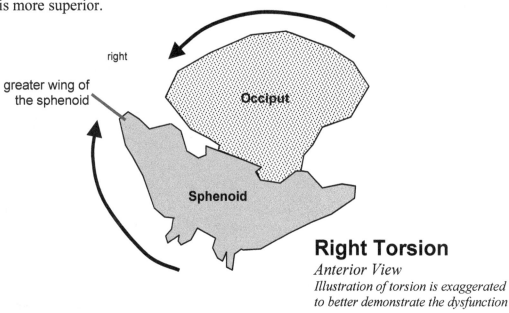

Right Torsion
Anterior View
Illustration of torsion is exaggerated to better demonstrate the dysfunction

SIDEBENDING/ROTATION - this involves two motions about three axes. Rotation occurs about an anterior-posterior (AP) axis that runs through the sphenobasilar synchondrosis (SBS), while sidebending occurs about 2 axes: 1 passes through the foramen magnum and the other through the sphenoid bone's center. For example, right sidebending will always be associated with spheno-occipital rotation that is superior on the left (SBS) deviated to the right). The sidebending is named for the direction in which the SBS moves (the "convex" side).

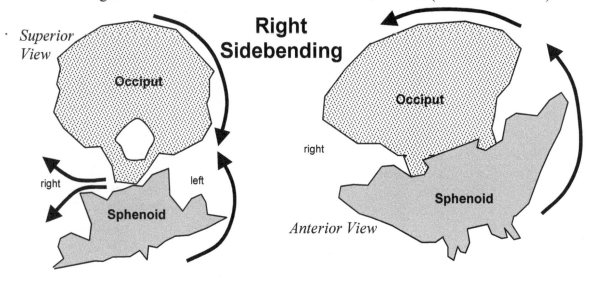

VERTICAL STRAIN - the sphenoid moves cephalad (superior vertical strain) or caudad (inferior vertical strain) with respect to the occiput. This movement occurs about two transverse axes - one through the sphenoid center and one just superior to the occiput.

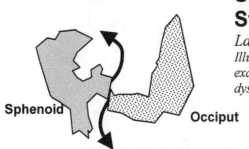

Superior Vertical Strain
Lateral View
Illustration of vertical strain is exaggerated to better demonstrate the dysfunction

LATERAL STRAIN - the sphenoid deviates laterally with respect to the occiput. The direction of the sphenoid deviation determines the type of lateral strain.

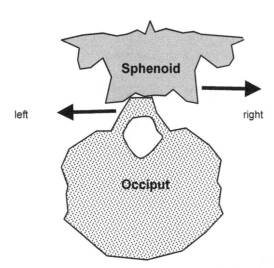

Right Lateral Strain
Superior View
Illustration of lateral strain is exaggerated to better demonstrate the dysfunction

COMPRESSION - sphenoid and occiput are compressed together. This usually occurs secondary to trauma to the back of the head. This prevents physiologic flexion and extension and, therefore, the CRI is absent.

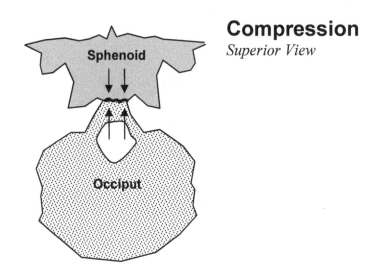

Compression
Superior View

According to Cranial Field theories, the following cranial nerves will be affected by the following cranial bones when the latter are in dysfunction:

SPHENOID	CN I, II, III, IV, V, VI, VII, VIII
FRONTAL	CN I
ETHMOID	CN I
OCCIPUT	CN II, VII, VIII, IX, X, XI, XII
TEMPORAL	CN III, IV, V (v1, v2), VI, VII, VIII, IX, X, XI

Other Bones/Joints to Consider:

Maxillae	V2 of CN V
Mandible	V3 of CN V
OA, AA, C2	CN X

As espoused by Sutherland. the goals of craniosacral treatment are:
1. Reduction of venous congestion
2. Mobilization of articular restrictions
3. Attainment of SBS balance
4. Enhancement of amplitude and rate of the CRI

Remember, all craniosacral techniques rely on the principle that the brain, spinal cord, and CSF are theoretically movable/moving, and that the dura has key attachments throughout the CNS.

Some Important Craniosacral Techniques

CV4: BULB DECOMPRESSION
This technique is utilized to increase the CRI amplitude by decompressing any SBS compression that may exist. To perform this technique, the flexive and extensive forces of the occiput are resisted by the practitioner until a "still point" is reached, after which the area is released to permit restoration of normal flexion and extension with the concomitantly enhanced CRI amplitude.

VAULT HOLD
This is used to modulate SBS strains by balancing membranous tensions. To perform this, the patient is placed in a supine position. The clinician places his thumbs in a manner that they cross over but do not touch the sagittal suture. The index fingers are placed over the greater wing of the sphenoid. The middle fingers are placed on the squamous portion of the temporal bone (the area just anterior to the ears). The ring fingers are placed over the mastoid process, and the little fingers are positioned over the squamous portion of the occiput. Then the finger pads are used to induce motion.

V SPREAD
This is used to release a suture that is restricted. It is a technique that requires physically disengaging the suture.

VENOUS SINUS RELEASE
This technique is used to release the venous sinuses and to encourage good venous drainage from the head. It is frequently employed to help relieve the symptoms of headache.

Indications for Craniosacral Treatment

1. Any SBS strain pattern
2. Head trauma, i.e. forceful dental procedures, assault, motor vehicle accidents
3. Newly delivered neonate to realign cranial bones that may have been traumatized during delivery; the goal is to prevent the development of cranial bones growing together to form a synostosis.

Complications of Craniosacral Techniques
1. Headache
2. Tinnitus
3. Dizziness
4. Altered heart rate, blood pressure, and respiratory rate

Absolute Contraindications to Craniosacral Treatment

1.   Skull fracture
2.   Intracranial bleed of any type
3.   Increased intracranial pressures

Also, patients with CNS neurological disorders, particularly seizure disorders, as well as traumatic brain injury should be approached with caution.

The following scenarios allow for exercise in determining the theoretical dysfunction based upon palpatory and visual inspection of the patient's head.

So, what dysfunction exists if...

o The greater wings of the sphenoid are inferior and anterior bilaterally:
    Flexion
o The greater wing of the sphenoid is inferior on the right and superior on the left
    AND the occiput is inferior on the right and superior on the left:
    Right sidebending rotation
o The greater wing of the sphenoid is superior on the left and inferior on the right
    AND the occiput is inferior on the left and superior on the right:
    Left torsion
o The occiput is superior and medial bilaterally:
    Extension
o The right eye is receded and the left eye is prominent AND the right ear is
    prominent and the left ear is very close to the head: Right sidebending rotation
o The right eye is prominent and the left eye is receded AND the right ear is
    prominent and the left ear is very close to the head:
    Right torsion
o The eyes are prominent bilaterally and the ears are prominent bilaterally:
    Flexion
o The eyes are receded bilaterally and the ears are close to the head bilaterally:
    Extension
o The palate is high and narrow bilaterally and the orbits are narrow bilaterally:
    Extension
o The left temporal bone is externally rotated while the right temporal bone is
    internally rotated:
    Left sidebending rotation or Left torsion
o The left temporal bone is externally rotated while the right temporal bone is
    internally rotated AND the left mastoid is positioned posteriomedially:
    Left sidebending rotation or Left torsion
o The left temporal bones is externally rotated while the right temporal bone is
    internally rotated AND the left mastoid is positioned posteriomedially AND the left
    eye is receded:
    Left sidebending rotation
o The temporal bones are internally rotated bilaterally:
    Extension

Chapter 10 Review Cases

1. A patient is found to have externally rotated paired bones of the skull. What would be an expected associated finding?
    a. Narrow head
    b. Increased A-P diameter of the head
    c. Sacral nutation
    d. Lax reciprocal tension membrane
    e. Craniosacral flexion

2. Right sidebending of the spheno-occipital unit is associated with what?
    a. Rotation of the spheno-occipital unit is superior on the right around an A-P axis
    b. Left side occiput more superior than right side
    c. SBS deviation to the left
    d. Sphenoid rotation to the right around a vertical axis that passes through the sphenoid
    e. Movement about 4 axes

3. What SBS strain results in loss of the CRI?
    a. Inferior vertical strain
    b. Left lateral strain
    c. Craniosacral extension
    d. Compression
    e. Left cranial torsion

4. You identify a patient as having tinnitus. What cranial bone, besides the sphenoid and occiput, may be in dysfunction to cause this?
    a. Frontal bone
    b. Ethmoid bone
    c. Temporal bone
    d. Nasal bone
    e. Parietal bone

5. Upon examination of a 17-year old patient, you find that the greater wings of the sphenoid are anterior and inferior bilaterally. What dysfunction exists?
    a. Craniosacral flexion
    b. Left sidebending
    c. Craniosacral extension
    d. Compression
    e. Superior vertical strain

6. What findings would you expect with respect to the eye, the ear, and the mastoid in a patient with right sidebending rotation?
    a. Receded right eye, prominent left ear, and posteriomedial right mastoid
    b. Receded left eye, prominent left ear, and posteriomedial left mastoid
    c. Receded left eye, prominent left ear, and posteriomedial right mastoid
    d. Receded right eye, prominent right ear, and posteriomedial left mastoid
    e. Receded right eye, prominent right ear, and posteriomedial right mastoid

7. During physical examination of a patient's skull, it is determined that the greater wing of the sphenoid is superior on the left and inferior on the right, while the occipital bone is superior on the right and inferior on the left. What eye and ear findings are expected?
    a. Left eye and right ear are prominent
    b. Left eye and ear are prominent
    c. Right eye and left ear are prominent
    d. No anomalies are expected
    e. Right eye and ear are prominent

Answers to Chapter 10 Cases

1. E   External rotation of the paired bones of the skull is, according to craniosacral theories, associated with craniosacral flexion and is responsible for the wide head reported with such strain.

2. B   In assessing cranial field concepts, right sidebending is linked to the occiput, as well as the sphenoid, being superior on the side contralateral to the sidebending.

3. D   Compression reportedly results in decreased or absent cranial rhythm impulse (CRI).

4. C   Tinnitus may result from cranial nerve VIII dysfunction. Cranial nerve VIII may be disrupted by dysfunction of the sphenoid bone, the occipital bone, or the temporal bone.

5. A   According to these concepts, if the anterior portion of the sphenoid, palpated via the greater wings of sphenoid, is anterior and inferior, the anterior aspect of it must be dipping down, as if it itself is in flexion. This is consistent with craniosacral flexion.

6. E   If there is right sidebending, it is espoused that the SBS is deviated to the right. While the SBS is deviated to the right, it is also believed that the left side of both the occiput and sphenoid are deviated upward. This theoretically causes a receded right eye (ipsilateral to the sidebending), a prominent right ear (ipsilateral to the sidebending), and a posteriomedial right mastoid.

7. B   According to theories developed by Sutherland, if the greater wing of the sphenoid is superior on one side of the head while the occipital bone is found to sit more superiorly on the opposite side of the skull, a torsion between the sphenoid and occiput is created. This is called a cranial torsion, and is named for the side with the superior greater wing of the sphenoid; thus, in this case, there is a left cranial torsion. This hypothetically can create changes in the person's appearance such that the eye and ear ipsilateral to the torsion are both prominent. Since this is a left cranial torsion, this would then be expected, according to cranial theories, to produce a prominent left ear and eye.

APPENDIX A: Synopsis of Clinical Assessment Tests

Test	Element Assessed	Interpretation
Standing flexion test	Confirms side of somatic dysfunction of the innominates, sacrum, or lower extremity	During flexion while standing, the PSIS will ride up more on the side of dysfunction and the test will be interpreted as positive on that side
Seated flexion test	Confirms side of somatic dysfunction of the sacrum (and, according to some authorities, the innominates)	During flexion while seated, the PSIS will ride up more on the side of dysfunction and the test will be determined as positive on that side
Lumbosacral spring test	Identifies unilateral or bilateral posterior positioning of the sacral base	Poor quality springing over the lumbosacral junction renders the test positive, and indicates that there is either a unilateral or bilateral posterior sacral base
Innominate rocking test	Confirms the side of somatic dysfunction of the sacrum, innominates, or pubic symphysis	Also known as the ASIS Compression Test. Poor quality springing over the ASIS yields a positive test and identifies the side of somatic dysfunction.
Apley's scratch test	Tests range of motion of shoulder	Poor range of motion per the test standards yields an inappropriate response, and confirms presence of limited range of motion of the shoulder
Adson's test	Identifies presence of thoracic outlet syndrome	Decreased or absent radial pulse while arm is extended and abducted and head is turned toward that arm renders the test positive, identifying the likely presence of thoracic outlet syndrome
Roos test	Identifies presence of thoracic outlet syndrome	Also known as the Elevated Arm Stress Test While both arms are flexed at elbow, externally rotated, and abducted to 90 degrees, inability to continue repeated action clenching of fist renders test positive

Test	Element Assessed	Interpretation
Speed's test	Presence of bicipital tendinitis	While patient undertakes isotonic contraction in an effort to extend the elbow, supinate forearm, and flex the arm at elbow, tenderness at the bicipital groove during that time renders the test positive and indicates presence of bicipital tendinitis
Yergason's test	Identifies instability of biceps tendon in bicipital groove	The patient flexes the elbow flexed to 90 degrees, but has further flexive forces resisted by the clinician and, at the same time, resists external rotation of the forearm being induced by the clinician; pain at the bicipital groove or movement of the tendon out of the groove constitutes a positive test, and indicates the presence of instability of the biceps tendon (usually due to bicipital tendonitis)
Tinel's test	Identifies presence of carpal tunnel syndrome	If tapping over the volar aspect of the wrist produces pain or paresthesias on the palmar aspect of the thumb and first 2 ½ fingers, the test is rendered positive and is indicative of carpal tunnel syndrome
Phalen's test	Identifies presence of carpal tunnel syndrome	If wrist flexion produces pain or paresthesias of the palmar aspect of the thumb and first 2 ½ fingers, the test is rendered positive and is indicative of carpal tunnel syndrome
Prayer test	Identifies presence of carpal tunnel syndrome	If wrist extension produces pain or paresthesias of the palmar aspect of the thumb and first 2 ½ fingers, the test is rendered positive and is indicative of carpal tunnel syndrome
Sulcus sign	Identifies inferior shoulder instability (when test is positive)	If a sulcus or indentation is seen inferior or lateral to the acromion while the clinician pulls down on the arm, the test is positive

Test	Element Assessed	Interpretation
Anterior drawer test	Identifies anterior cruciate ligament tears	With knee flexed, if the lower leg moves anteriorly while the clinician pulls it anteriorly, the test is positive and indicative of an ACL tear
Posterior drawer test	Identifies posterior cruciate ligament tears	With knee flexed, if the lower leg moves posteriorly as the clinician pushes against it, the test is positive and indicative of a PCL tear
Apley's compression test	Identifies meniscal damage to the knee	With the knee flexed and the patient prone, the clinician pushes down on the foot, compressing the knee joint while also internally and externally rotating the tibia. Pain renders the test positive, and the side of pain identifies the meniscus that is injured
Apley's distraction test	Identifies injury to the collateral ligaments of the knee	With the knee flexed and the patient prone, the clinician pulls up on the foot, distracting the knee joint while also internally and externally rotating the tibia. Pain renders the test positive, and the side of pain identifies the side of collateral ligamentous damage
Lachman's test	Identifies anterior cruciate ligament tears	Similar to an anterior drawer test, but with the patient supine and the knee up and braced against the clinician's body; movement of the tibia anteriorly renders the test positive and indicates an ACL tear
McMurray's test	Identifies tears of the posterior aspect of the menisci of the knee	Through flexion, either valgus or varus stress of the knee, and internal or external rotation, a click with particular positions confirms the presence and side of a tear
Patellar femoral grinding test	Identifies pathology of the posterior patellar surface or trochlear groove	Crepitance while providing downward resistance to the patella as the patient flexes their quadriceps renders the test (+)

Test	Element Assessed	Interpretation
Chapman's reflex	Identifies source of viscerosomatic-induced pain	Pain induced with palpation over well-described somatic locations, known as Chapman's points, can render a positive identification of such a point, and can be suggestive of the visceral source of the somatic pain being elicited. Once the source is identified, the visceral structure can be assessed for pathology.
Travell's myofascial trigger point	Identifies somatosomatic, somato-visceral, and, occasionally, viscera-somatic reflexes	Pain or irritation on the body that generates pain or an autonomic response elsewhere in the body, particularly when the original inciting painful or irritated area is palpated, is diagnostic for a trigger point, rendering the finding positive and identifying the possibility of facilitation that must be addressed
Jones tender point	Identifies dysfunction of myofascial structures	Palpation of myofascial structures that generates focal pain over the area of palpation renders the test positive for a Jones tender point, and is often suggestive of somatic dysfunction of the underlying myofascial structure
Red reflex test	Identifies an area with probable acute somatic dysfunction	Excessive and persistent reddening of the skin after application of a mild friction force renders the test positive, and suggests possible acute somatic dysfunction in that area.
Deep tendon reflex test	Identifies dysfunction of the peripheral nervous system or the central nervous system	Generally, increased DTRs are indicative of upper motor neuron dysfunction (e.g., motor neurons in CNS) while decreased DTRs are indicative of lower motor neuron dysfunction (e.g., nerve roots and their distal branches)
Vault Hold	Identifies cranial bone dysfunction	While this is also a treatment modality, the positioning can also used to evaluate the cranial bones

Test	Element Assessed	Interpretation
Spurling's test	Identifies cervical radiculitis/cervical nerve root compression	Also known as the Foraminal Compression Test If pain radiates down the arm on the same side to which the patient is, during the test, sidebending the head and concomitantly having the clinician push downward on the head, the test is rendered positive. The pain is usually dermatomally distributed, thereby also allowing for identification of which nerve root is being irritated or impinged.
Cervical distraction test	Identifies cervical radiculitis/cervical nerve root compression	If cervical or upper extremity pain is relieved by the clinician lifting the head up from under the chin and occiput, the test is considered positive and indicative of cervical nerve root pathology
Romberg's test	Assesses integrity of proprioception	If patient loses their balance while standing for 20-30 seconds with feet together and eyes closed, the test is positive and suggestive of proprioception problems. It may also be indicative of vestibular problems or even cerebellar pathology (although the latter will produce the same results regardless of the eyes being open or closed)
Underberg's test	Indentifies poor perfusion to brain, and may be suggestive of stenosis of the vertebral, basilar, or carotid arteries	The patient marches in place while arms are flexed to 90 degrees, hands supinated, eyes closed, neck extended, and head rotated to one side, and then the other side. Failure to keep the arms up, to maintain balance, or to retain supination of the hands renders the test positive.
Crank test	Identifies chronic shoulder dislocation	Also known as the Apprehension Test; If the patient purposefully resists any further movement while the clinician slowly externally rotates the abducted arm, this is considered positive.

Test	Element Assessed	Interpretation
Empty can test	Identifies presence of tears in the supraspinatous muscle or tendon	Pain or weakness experienced as the patient, having the arm abducted to 90 degrees, moves the arm forward about 30 degrees and internally rotates it as if to pour from a can is definitive of a positive test. Occasionally, a positive test can also be found with neuropathy of the suprascapular nerve.
Drop Arm Test	Identifies tears to the rotator cuff muscles or their tendons, particularly the supraspinatous structures	If pain or inability to drop the arm slowly occurs as the patient tries to slowly lower the fully abducted arm, the test is considered positive. Positive tests are also often obtained with supraspinatous tendinitis.
Lasegue's test	Identifies lumbar nerve root pathology (usually secondary to disc herniation)	Also known as the Straight Leg Raising Test. While the patient is supine, the leg at the hip is flexed to beyond 35 degrees until pain, if any, is felt. Pain in the low back accompanied by pain down the leg in a radicular pattern is indicative of a positive test. It should be followed by the Bragard's test to increase the accuracy and specificity of the findings.
Hip drop test	Assesses sidebending capacity of the lumbar and thoracolumbar spine	If the hip drops less than 20-25 degrees while the standing patient flexes one knee without lifting the heel off the floor, the test is considered positive and reflective of poor sidebending capacity. This is often used in evaluating the severity of spinal disorders, including scoliosis. DO NOT confuse this with the Trendelenburg test, a completely different assessment!
Trendelenburg's test	Assesses strength of gluteus medius muscle and integrity of superior gluteal nerve	If the hip drops at all while the patient is standing and lifting the ipsilateral leg OFF the floor, the test is positive for pathology

Test	Element Assessed	Interpretation
Bragard's test	Confirmation of lumbar nerve root pathology (usually secondary to disc herniation)	This is done immediately after the Lasegue's test, and confirms the former test's findings. After pain is elicited in the Lasegue's, the leg is lowered slightly to the position that relieves the pain. Then, the foot is dorsiflexed. If the pain produced during the Lasegue's test is reproduced, the Bragard's test is positive and provides confirmation of the Lasegue's test findings.
Hoover test	Detects possible malingering	If while patient is supine, the patient has "difficulty" raising a leg on request but fails to create any downward pressure through the contralateral heel, the test is positive and the patient may be malingering.
Patrick's test	Screens for hip joint pathology	Also known as the FABER or FABERE test; there are two ways to do it. With patient is supine, the leg is flexed and the foot is placed on the opposite knee after which the knee of that same leg is lowered to cause ipsilateral hip abduction. Failure to lower the knee to a level horizontal to the level of the opposite leg renders the test positive, and indicates hip pathology. FABER stands for Flexion, ABduction, and External Rotation.
Finkelstein test	Tenosynovitis of the abductor pollicis longus and the extensor pollicis brevis (DeQuervain's tenosynovitis)	If pain is experienced while the patient makes a fist with thumb tucked inside and, at the same time, adducts the wrist, the test is positive and identifies this tenosynovitis

Handbook of OMT Review

OMM Review Cases

1. While evaluating a 32-year old female patient in neutral position, you notice that your left thumb is more posterior than the right when assessing rotation of L3. Symmetry is restored in flexion. What is the dysfunction?
   a. L3FRLSL
   b. L3ERLSL
   c. L3NSRRL
   d. L3NSLRL
   e. L3FRRSR

2. A teenager is riding a horse, and suddenly and forcefully pulls back on the reigns since a vehicle has driven right in front of her and her horse. What kind of contraction did she undertake to execute this movement?
   a. Isotonic, eccentric
   b. Isometric, concentric
   c. Isolytic
   d. Eccentric
   e. Concentric

3. T3 is stuck sidebent left when in flexion. Into what position will you place this patient in order to perform an indirect technique?
   a. T3 neutral, rotated right, sidebent left
   b. T3 flexed, rotated left, sidebent left
   c. T3 flexed, rotated right, sidebent right
   d. T3 neutral, rotated left, sidebent right
   e. T3 extension, rotated right, sidebent right

4. An infant is born with a neck that is grossly sidebent left with the chin pointing to the right shoulder. What is this?
   a. Down's syndrome
   b. Spasmodic torticollis due to right sternocleidomastoid spasm
   c. Spina bifida
   d. Congenital torticollis due to left sternocleidomastoid contracture
   e. Cerebral palsy

5. A tender point is appreciated posterior to the clavicles, immediately at the base of the neck. What other somatic dysfunction do you expect to discover?
   a. Tonsillitis
   b. First exhalation rib
   c. Second inhalation rib
   d. Herniation of the T1 intervertebral disc
   e. Hepatitis

6. If a patient has lymphatic congestion of the heart, what vessel obstruction may have been the cause?
   a. Right lymphatic duct
   b. Left duct
   c. Inferior vena cava
   d. Thoracic duct
   e. Cysterna chili

7. As a patient moves his right hand into dorsiflexion, what is the plane and axis, respectively?
   a. Coronal plane; transverse axis
   b. Transverse plane; vertical axis
   c. Horizontal plane; A-P axis
   d. Anterior-posterior plane; vertical axis
   e. Sagittal plane; transverse axis

8. You treat a patient, using traction during myofascial release as the patient fully relaxes. How would you describe the method used?
   a. Passive indirect
   b. Passive direct
   c. Active indirect
   d. Active direct
   e. Active passive

9. While evaluating a patient, you find that T4-T6 cannot sidebend to the right when the patient is in neutral; sidebending is restored in flexion. What is the restriction?
   a. T4-T6FRLSL
   b. T4-T6ESLRL
   c. T4-T6NSLRR
   d. T4-T6FSRRL
   e. T4-T6NRRSR

10. A 49-year old patient with recurrent abdominal pain and chronic nausea presents for evaluation. During palpation, you identify a tender point on the right 7th costochondral joint. When you palpate the epigastric area, the patient complains of shooting pain in the skin of the right lower chest wall. What will you include in your differential?
    a. Gastritis
    b. RLL pneumonia
    c. Biliary disease
    d. Chronic pancreatitis
    e. Hypersplenism

11. A 22-year old male presents with complaints of a flexion contracture of the middle portion of the deltoid. How would you perform post-isometric relaxation?
    a. Adduct arm at shoulder; then have patient adduct arm more against resistance
    b. Adduct arm at shoulder; then have patient abduct arm against resistance
    c. Abduct arm at shoulder; then have patient abduct arm against resistance
    d. Abduct arm at shoulder; then have patient adduct arm against resistance
    e. Flexion at the shoulder; then have patient extend arm against resistance

12. While evaluating a patient, you find that C3 is stuck translated right when flexed. What is the diagnosis?
    a. C3FRLSL
    b. C3FRRSR
    c. C3ERRSR
    d. C3NRLSL
    e. C3NRRSL

13. In evaluating a patient, you note that her left eye is receded and her left ear is prominent. What craniosacral dysfunction exists?
    a. Compression
    b. Right sidebending rotation
    c. Flexion
    d. Right torsion
    e. Left sidebending rotation

14. Approximately half of cervical flexion occurs at the
    a. OA
    b. AA
    c. C2-C3 division
    d. Inferior division of the cervical spine and AA
    e. C7-T1 joint

15. A 53-year old patient fell down the stairs at his home last week. He presents today with complaints of pain and paresthesias in his medial right hand. On palpation of the cervical spine, cervical motion tenderness is demonstrated. What disc could have herniated to generate these results?
    a. C4
    b. C5
    c. C6
    d. C7
    e. C8

16. A middle-aged patient presents with complaints of a sensation of not being able to breathe normally as he feels as if he cannot move his ribcage fully. During evaluation, you appreciate that the superior edge of the posterior rib angle of rib 3 is prominent. What is the dysfunction that would account for this patient's symptoms?
    a. Inhalation restriction
    b. Exhalation dysfunction
    c. Exhalation restriction of a false rib
    d. Inhalation rib
    e. Group inhalation dysfunction

17. On evaluation of rib 8, the lateral part of the shaft is determined to be elevated. What restriction exists?
    a. Exhalation dysfunction
    b. Inhalation dysfunction
    c. Exhalation restriction
    d. Inhalation restriction
    e. No dysfunction exists

18. In assessing a patient's spine, you find that the left transverse process of the T3 segment is more prominent than the right, as is most noticeable in flexion and somewhat noticeable in extension. If you treat this with post-isometric relaxation, what movement will you ask him to try to execute after you set him up for the treatment?
    a. Rotate right and sidebend left
    b. Rotate left and sidebend right
    c. Rotate left and sidebend left
    d. Rotate right and sidebent right
    e. Flex and rotated left

19. A 45-year old G4P3 is entering labor. She is currently suffering from diverticulitis of the sigmoid colon. What might be an expected outcome?
    a. Prolonged labor
    b. Rapid delivery of the infant
    c. Chorioamnionitis
    d. Neonatal meningitis with GBS
    e. An almost painless delivery process

20. An infant is born with paralysis of the lower extremities. Imaging reveals failure of closure of the laminae of the vertebral arch of several vertebrae. What is the most likely diagnosis?
    a. Congenital hydrocephalus
    b. Cerebral palsy
    c. Spina bifida occulta
    d. Spina bifida meningocele
    e. Spina bifida meningomyelocele

21. A 61-year old construction worker injures himself while trying to pick up a bucket of cement. He presents with complaints of acute onset, severe, sharp low back pain that is worsened by forward bending. Physical exam reveals a decreased right patellar reflex and hyperesthesia of the medial leg between the knee and the ankle. What is the diagnosis?
    a. Spinal stenosis of the L3-L5 vertebral segments
    b. Herniation of the L4 intervertebral disc
    c. Traumatic spondylolisthesis
    d. Spondylolysis
    e. L3 nucleus pulposus herniation

22. If a patient has a T4ERRSR dysfunction, where will you place your thenar eminence and to what side will you sidebend the patient before applying the thrust?
    a. Under the posterior transverse process of the dysfunctional segment; sidebend right, respectively
    b. Under the posterior transverse process of the dysfunctional segment; sidebend left, respectively
    c. Under the anterior transverse process of the dysfunctional segment; sidebend right, respectively
    d. Under the anterior transverse process of the vertebra below the dysfunctional segment; sidebend right, respectively
    e. Under the posterior transverse process of the vertebrae below the dysfunctional segment; sidebend left, respectively

23. You wish to treat a C4FRRSR with HVLA. How will you set up this patient, and in what direction will you incur the thrust?
    a. Slight flexion of C3, head and neck rotated right and sidebent right; thrust rightward directed towards the positioning of the patient's eye
    b. Slight anterior translation at C3, head and neck rotated right and sidebent right; thrust rightward directed towards the positioning of the patient's eye
    c. Slight anterior translation to C3, head and neck rotated left and sidebent left; thrust leftward directed towards the position of the patient's eye
    d. Slight flexion of C3, head and neck rotated left and sidebent left; thrust leftward directed towards the position of the patient's eye
    e. Slight anterior translation at C3, head and neck rotated left and sidebent left; thrust rightward directed towards the position of the patient's eye

24. A 33-year old female patient is diagnosed with ulcerative colitis. What other findings might you expect?
    a. Hyperovulation
    b. Decreased glomerular filtration rate
    c. Increased vitamin B12 absorption
    d. Menorrhagia
    e. Decreased potential for urinary tract infections

25. During palpation of a patient's skull, you note that your index fingers placed over the sphenoid are more inferior and anterior than usual. Your digit minimi over the occiput are inferiorly and laterally displaced. What presenting complaint of this patient would have been consistent with these findings?
    a. Tinnitus
    b. Prominent left ear
    c. Bilaterally receded eyes
    d. Narrow upper jaw
    e. Beetle brow

26. A young athlete presents with complaints of transient pains in his dorsal right foot. He denies any back trauma, and no pain is felt on palpation to the back during his examination. He also complains that, while he sleeps, his foot annoyingly always ends up being "pointed" and pigeon-toed. What diagnosis could account for these complaints?
    a. Patella-femoral tracking syndrome
    b. Ankle sprain
    c. Congenital pronation of the foot, exacerbated by athletics
    d. Posterior fibular head
    e. Plantar fasciitis, exacerbated by poor footwear doing athletics

27. A 25-year old patient presents with complaints of low back pain after having accidentally stepped into a hole in his yard. On palpation, you find that the right ASIS, right PSIS, and the right pubic bone are superior. Pain is produced when the right sacroiliac joint is palpated. The standing flexion test is positive contralaterally. What is the diagnosis?
    a. Superior right pubic shear
    b. Superior right innominate shear
    c. Superior left innominate shear
    d. Inferior left innominate shear
    e. Inferior left pubic shear

28. A 68-year old male is getting a full osteopathic examination. His right superior sulcus is deep, his left superior sulcus is slightly shallow, and his left inferior lateral angle is shallow. He admits to having low back pain exacerbated by backward bending. The seated flexion test is positive on the right. What lumbar finding do you expect?
    a. L5NSLRR
    b. L5 extended
    c. No dysfunction of L5
    d. L5NSRRL
    e. L5ERRSR

29. During a sphinx test, you note that the superior sacral sulci become less asymmetric. What is the diagnosis?
    a. Forward sacral torsion
    b. Backward sacral torsion
    c. Bilateral sacral flexion
    d. Bilateral sacral extension
    e. There is no diagnosis as this is not consistent with the presence of a dysfunction, and is instead physiological

30. A patient who just returned from a 12 hour car ride presents with complaints of low back pain plus sciatic pain through the right buttocks and posterior thigh. During evaluation, you find a right pelvic shift and a left sacral oblique axis. What lumbar findings do you expect?
    a. Type I Fryette dysfunction
    b. Non-neutral dysfunction with rotation and sidebending to the left side
    c. Neutral dysfunction with a right rotational component
    d. Extension dysfunction with rotation and sidebending to the right side
    e. Spondylolisthesis, low grade, in conjunction with a compensatory small Ferguson's angle

Answers to OMM Review Cases

1. **A**   If the thumb is more posterior on the left than the right when evaluating rotation, that means the segment is rotated left. Thus, the rotational freedom – what it does with the most ease – is rotation left. However, the position in the sagittal plane that affords the freedom for the spine is flexion. After all, it is flexion that restores symmetry, thereby moving the spine out of a locked-up, dysfunctional position and placing it into a state of freedom. Thus, the dysfunction (freedom) is flexion. Since this a lumbar vertebral segment, Fryette principles must be applied. If it is flexed, rotation and sidebending must be to the same side. Thus, if it is rotated left, it must also be sidebent left. Therefore, it is L3FRLSL.

2. **E**   By pulling back on the reigns, she shortened the muscles of her arm, allowing her to pull back and create tension on the reigns. Any muscle contraction that results in muscle shortening is concentric contraction. It was not isotonic because, when she started to pull, there probably was little to no tension on the reigns; and then, when the reigns became taut, she had to likely apply greater force. Contraction can only be considered isotonic if the amount of tension applied during the contraction is constant; it likely was not herein. Thus, it is not isometric since there was movement due to the contraction.

3. **A**   If a segment is "stuck" in a position, that is the only position it can undertake – it is restricted from assuming any other position. Thus, its freedom is that position into which it is stuck. Thus, T3 is sidebent left. The freedom in the sagittal plane (flexion, extension, neutral) will be that position that affords the most freedom to the segment/spine. Flexion promotes T3 being in dysfunction, namely "stuck" sidebent left. Therefore, flexion is not the freedom. The freedom has to be neutral or extension. As no more detail is provided, we cannot ascertain which it is. Thus, the freedom or dysfunction for T3 is NSLRR or ERLSL. An indirect technique is any technique in which the dysfunctional tissue is moved away from its restrictive barrier. So, it is moved away from the position of restriction. To move something away from its restriction is the same thing as moving it towards its freedom. Thus, an indirect technique is one in which the dysfunctional tissues are placed into their position of freedom. The dysfunction is named for the positions of freedom. Since the dysfunction is either T3NSLRR or T3ERLSL, the patient is either placed into T3 neutral, sidebent left, rotated right position or T3 extended rotated left, sidebent left position. Only one correct option is offered among the options: T3 neutral, sidebent left, rotated right.

4. **D**   Torticollis can be acquired ("spasmodic torticollis") or it can be congenital. Congenital is a form of torticollis with which one is born, and is colloquially known as wryneck. This infant has congenital torticollis. Torticollis is due to spasm or contracture of the sternocleidomastoid muscle (SCM). Spasm or contracture of the SCM causes contralateral rotation and ipsilateral sidebending. This child has left sidebending. His chin is pointing to his right shoulder, so he is also rotated right. Thus, the dysfunctional SCM is on the left.

5. C   A tender point posterior to the clavicles at the base of the neck can be a
tender point within the scalenes. This indicates that there may be somatic
dysfunction of the scalenes. Such dysfunction can feature contracture and/or spasm
of the scalenes. Scalene contraction can result in lifting up of ribs to which the
scalenes are attached. The anterior and middle scalenes can, therefore, cause a first
inhalation rib, and a posterior scalene can cause a second inhalation rib. Exhalation
rib is not possible since the muscles will not push the bones down. While tender
points that are Chapman's points can be associated directly with the clavicle, those
tender points are not generally posterior to the clavicle and they do not signify
somatic dysfunction, as this question requested. Chapman's are used to identify
visceral dysfunction. Herniation of the T1 intervertebral disc would produce
symptoms associated with T1 nerve root, which does not innervate the scalenes or
nearby structures, and the dermatomes served by it are only on the upper medial
arm and a thin strip that extends across the upper back and chest.

6. D   The right arm, right side of the head and face, and the right thorax drain into
the right lymphatic duct. The rest of the body, including the heart, drains into the
thoracic duct.

7. E   Any questions like this, unless otherwise specified, refer to the patient in the
anatomic position. If a patient is in anatomic position, and dorsiflexes the hand, the
axis is transverse. Remember, like with any movement around an axis, the hand is
moving about an axis like a door around its hinges. In anatomic position, that axis
is a transverse axis through the wrist. The plane through which it moves is the
sagittal plane. After all, in anatomic position, when the hand flexes or dorsiflexes, it
is using the same axis and plane as torso flexion and extension.

8. B   In any treatment regimen in which the patient relaxes and does not <u>actively</u>
participate, the technique is said to be passive. Likewise, when moving the
myofascial structures away from the restrictive barrier, the tissues ultimately are
compressed to some extent. This is an indirect technique. However, in order to
establish direct treatment with myofascial release, one must move the myofascial
structures into the restrictive barrier. This requires placing traction on the
myofascial structures – the traction provides the force to move it into and through
the restriction (the barrier).

9. E   If T4-T6 cannot sidebend to the right when the patient is in neutral, that implies that, in neutral, it sidebends preferentially to the left. Therefore, the freedom (dysfunction) is sidebent left. The freedom in the sagittal plane (flexion, extension, neutral) is flexion because that is when full vertebral freedom is restored, since sidebending freedoms are restored in flexion, as noted in the case. So, the freedom (dysfunction) is flexion. Since the patient's dysfunction is T4-T6FSL, by applying Fryette mechanics, we know that the patient is rotated left. Thus, if the freedom is T4-T6FRLSL, we know the restriction (what the question asks for) is the opposite of that: T4-T6NRRSR. We know it is neutral rather than extension because it is neutral that established the sidebending restriction in the case. Also, remember that restriction does not follow Fryette mechanics; restriction is simply the opposite of the freedom – so that is why, <u>for restriction</u>, it is possible to have rotation and sidebending to the same side.

10. D   When pain is produced on palpation of the right 7th costochondral joint, that is a feature of a Chapman's point; the pain elicited through viscero-somatic reflex at that location indicates possible pathology of the pancreas. When one palpates a part of the body, and that palpation produces pain elsewhere, that is a Travell's trigger point. Trigger points represent the presence of reflex arcs. If palpating the epigastrium, one may be palpating the stomach or pancreas. If that palpation causes referred pain to the skin of the right lower chest wall, it is referring pain to the T7 dermatome. Thus, the origin must be from T7 stimulation. Either the stomach or the pancreas could be the source (both are sympathetically innervated by T7); however, the stomach is more likely to produce left-sided referred pain and the pancreas right-sided pain because the stomach's sympathetic innervation arises from the left and the pancreas' from the right. So, both the Chapman's point and the Travell's trigger point suggest pancreatic pathology. The best choice of the options provided is chronic pancreatitis.

11. B   The middle portion of the deltoid muscle is the primary abductor of the arm. If in contracture (chronic contraction), the arm is abducted. Post-isometric relaxation is a type of muscle energy. Muscle energy is typically a direct technique; thus, the patient must be placed into his restrictive barrier. If he is abducted via contracture, he is restricted from having his arm by his side (adducted). Thus, this patient's arm must be placed into his restrictive barrier: adduction. For post-isometric relaxation, after the set-up, the patient then must contract his muscle against resistance (so an isometric contraction is produced). The muscle responsible for the dysfunction is the one that is contracted. Thus, in this case, the middle portion of the deltoid must be contracted against resistance, thereby producing an isometric contraction. Therefore, after the patient is placed into adduction, the arm is then isometrically contracted towards abduction.

12. D   If C3 is translated right, it means it is sidebent left. If it is stuck in that position, that is the position it takes on more freely, since the opposite is impossible – and, thus, the opposite is the restriction. So, C3 is sidebent left since that is its "freedom." Flexion, however, promotes this dysfunction. So, in the sagittal plane, freedom is gained through either extension or neutral. Since this is C3, cervical motion rules apply. C3 has rotation and sidebending to the same side. Thus, the dysfunction is either C3ERLSL or C3NRLSL. C3NRLSL is the only option provided, so that is the correct answer. Remember, cervical spine does not follow Fryette mechanics, so it is possible to have a neutral dysfunction wherein the rotation and sidebending components are to the same side.

13. E   A receded left eye can imply that the left side of the sphenoid is lower than normal; such a finding is consistent with right torsion or left sidebending rotation. A prominent left ear can be created when the bulk of the occiput (not the SBS) is farther away from the left side of the head, leaving the ear, which overlies the temporal bone, to appear more isolated and separated from the head. That is consistent with left torsion or left sidebending rotation. Thus, the dysfunction that could account for both of these findings is left sidebending rotation.

14. A   About half of flexion of the cervical spine occurs at the level of the occipitoatlantal (OA) joint. About half of rotation of the cervical spine occurs at the atlantoaxial (AA) joint. The inferior division of the cervical spine includes vertebral segments C2-C7; they contribute the other 50% of flexion (and extension) and the other 50% of rotation to the cervical spine. Clinically (e.g., what we can palpate), AA does not undergo significant flexion or extension. So, in terms of flexion capacity, ½ comes from the OA and the other ½ from the inferior division. Nonetheless, we know from radiographic studies that the AA does undertake movements beyond just rotation. Thus, for academic purposes and movement that occurs below the palpatory threshold, some flexion and extension, and insignificant amounts of sidebending, do occur at the AA.

15. D   Pain and paresthesias anywhere in the body can have many causes, one of which being a herniated intervertebral disc. If the complaint involves the medial hand (remember, all things are always referred to as if the patient were in anatomic position), that means that the C8 dermatome may be involved. In other words, impingement of the C8 nerve root could do this. C8 nerve root can be impinged by a C7 intervertebral disc herniation. Thus, such symptoms could have been produced by a herniation of the C7 disc.

16. D   The posterior rib angle is found on the posterior aspect of a rib. The superior edge is the top of that rib. So, if the top edge of the posterior aspect of the rib (the part of the rib closest to the spine) is prominent or sticking out, that means that the top is tilted outward and the bottom is tilted inward; thus, that means that the back of the rib is tilted upward. If the back of the rib is tilted upward, this means that the front of the rib or the anterior aspect of the rib is pointing upward. This is rib 3, so it undertakes primarily pump-handle motion. Therefore, if the back is tilting upward, that definitely makes the front tilt upward. Rib dysfunction of pump-handle ribs is named for what the anterior aspect (e.g., tip of the shaft) is doing. So, if the anterior portion is upward, the rib is in a position that mimics that which it undertakes during inhalation. However, since the rib is stuck in this position regardless of the respiration cycle, it is in inhalation dysfunction. Inhalation dysfunction is synonymous with an inhalation rib and with an exhalation restriction (the latter meaning restricted from moving into a position consistent with exhalation).

17. C   The lateral shaft is that part of the rib at the side of the thorax. Herein, it is elevated. Rib 8 is a rib that moves primarily through bucket-handle movement. Thus, any dysfunction is named for what the apex of that handle is doing, specifically the lateral shaft. If the lateral shaft is elevated, the rib is elevated. If it is elevated, it is in a position that mimics that associated with inhalation. Since this position is not physiologic (occurring with only inhalation), but rather is constant, it is a dysfunction, namely an inhalation dysfunction. The question asks for the restriction. If it is an inhalation dysfunction, and dysfunctions are named for the freedom of movement, then the restriction is the opposite of the freedom – and, thus, the opposite of the dysfunction. Therefore, this rib is in exhalation restriction, meaning it is restricted from taking on an exhalation position.

18. B   If the left transverse process of T3 is more prominent than the right, that means that the segment is rotated left. This finding is most prominent in flexion and still somewhat noticeable in extension. Thus, neutral is the only position that affords the most freedom to this segment, and allows it to assume a position in which it is not "locked up" into one particular rotational dysfunction. So, neutral is the freedom. Thus, this is T3NRL. Applying Fryette mechanics, we know that this is T3NSRRL. If we are going to do post-isometric relaxation, that means we are doing muscle energy. Muscle energy is almost always a direct technique; so, the set-up is that of putting the patient into his restrictive barrier. If the freedom is T3NSRRL, the restriction is the opposite of that: T3FSLRR. We know flexion is its restriction because, in the case, we learned that flexion is the position that most forces T3 into its dysfunctional, non-asymmetric position. If the freedom is sidebending right and rotating left, then the restriction has to be sidebending left and rotating right. Remember, restriction follows one rule: it is the opposite of freedom, so Fryette does not apply to restriction. If the restriction is T3FSLRR, that is the position into which we set the patient up for muscle energy. However, the question asks about what movement we will ask the patient to try to execute (which we will control so that it is only an isometric contraction). For post-isometric relaxation, after the set-up, we have the patient contract the muscle or muscles responsible for the dysfunction. If the dysfunction is T3NSRRL, then it is the muscles that are maintaining or promoting that position that are the culprits. Thus, if we ask the patient to contract or move (although we allow no movement) in the direction of the dysfunction, we are essentially having the patient engage and contract the muscles responsible for that dysfunction. So, after the set-up, we ask the patient to sidebend right and rotate left.

19. A   The patient has diverticulitis, which is plugging and resulting infection of diverticuli. This is located in this patient to the sigmoid colon. Thus, the sigmoid is acutely diseased. Pathology, infection, trauma, and other anomalies launch the body's fight or flight mechanism: the sympathetic nervous system. Thus, disease in the colon can activate the sympathetics that serve the sigmoid colon. The sigmoid is served by T12-L2, sympathetically. Inappropriate afferent impulses can be delivered from the sigmoid to the T12-L2 spinal cord segments. With chronic stimulation or, as in this case, intense acute stimulation, the T12-L2 spinal cord segments can become facilitated, meaning that the incoming stimulation results in a decrease in those segments' threshold to activation. Such a decrease in threshold to activation (facilitation) allows for stimulation of other structures innervated by T12-L2. Motor neurons can be stimulated to cause hypertonicity and even spasm of muscles innervated by T12-L2, pain fibers associated with the T12-L2 segments can be stimulated to cause referred pain in T12-L2 dermatomes and structures with sensory innervation by those segments, and sympathetics arising from T12-L2 can deliver sympathetic stimulation to tissues innervated by T12-L2. The uterus is innervated sympathetically by T12-L2. Thus, the uterus could be inappropriately sympathetically stimulated, causing it to respond in a fight or flight manner, a response that is geared for immediate but not necessarily long-term survival. Thus, if this woman were really in a life-threatening, fight or flight situation (such as under attack by a wild animal, for instance), would it be more advantageous to survival of mother and infant to deliver now or to not deliver? Delivery "now" would slow her up, and also would expose the baby – putting both at risk. Not delivering is more advantageous to immediate survival. Thus, her current case of diverticulitis could result in prolonged labor as the uterus becomes flaccid and non-contractile during sympathetic stimulation. Likewise, she would potentially also have referred pain, adding to the pain of the delivery process.

20. E   Failure to have closure of the laminae of the vertebral arch is, by definition, spina bifida. What type of spina bifida is most likely to cause neurologic dysfunction, such as paralysis of the limbs? The kind that has allowed for herniation of nerve roots (in lumbar spine or sacrum) or spinal cord (in thoracic and cervical spine) out of the vertebral canal, thereby eliminating any protection from damage or pressure on those neurologic structures. That type of spina bifida is spina bifida meningomyelocele, also known as spina bifida myelomeningocele. Cerebral palsy could cause severe hypo- or, more commonly, hypertonia of all muscles of the body; a small head, and/or scoliosis, and/or small mandible are common features also at birth. Congenital hydrocephalus does not cause paralysis, but rather manifests in the infant as vomiting, lethargy, poor feeding, and a larger-than-normal head. Imaging reveals enlarged ventricles.

21. E   If he was picking something up, there is a good chance he was forward bending. Spinal flexion predisposes the lumbar intervertebral segments to herniation, especially when axial loading (e.g., the weight of the bucket and its contents) is placed upon the spine. This patient complains of sudden onset sharp pain, features consistent with a herniation of any tissue, including intervertebral discs. Also, exacerbation of that pain with flexion is consistent with a herniated disc. A decreased patellar reflex indicates pathology to the L4 nerve roots or any of its more distal branches. Sensory anomalies in the region of the medial leg between the knee and the ankle is consistent with pathology of the L4 dermatome. Thus, these findings are consistent with a compression of the L4 nerve root. If this is due to disc herniation, which is supported by the presenting complaints, then it is herniation of the L3 disc that would be responsible. When intervertebral discs herniate, it is the soft nucleus pulposus normally contained within the fibrous anulus fibrosus that is what is herniated out of the disc. So, an L3 nucleus pulposus herniation would account for this patient's problems.

22. E   The patient has a T4ERRSR dysfunction. Thus, his freedom of motion is that of T4 extended, rotated right, and sidebent right. HVLA applied to the thoracic spine typically is directed to the dysfunctional segment in flexion dysfunctions, and to the vertebra BELOW the dysfunctional segment in extension dysfunctions. Thus, since this is an extension dysfunction, treatment will be directed to the vertebra below the dysfunctional vertebra; specifically, the thenar eminence of your hand is placed beneath the vertebra below the dysfunctional segment. HVLA is a direct technique, which means that the patient is placed into his restrictive barrier. The restrictive barrier is the restriction. If T4 is rotated right, its freedom is rotating right and its restriction is rotating left. Thus, HVLA seeks to position the segment into its restrictive barrier: rotation left. But, remember, treatment is directed to the vertebrae below the dysfunctional segment. So, the thenar eminence is placed under T5, but under the left transverse process. That encourages, during the thrust, rightward rotation of T5; because neighboring body regions move in opposite direction, that will cause T4 to rotate left – into its restrictive barrier. That is part of the set-up; during the thrust, it should be noted that the thrust is directed cephalad (upward, so as to treat T4), specifically 45 degrees cephalad. As for the sidebending, which is also part of the set-up, remember that HVLA is a direct technique. So the patient must be placed into their restrictive barrier. If sidebending right is the freedom, then restriction is sidebending left. Thus, the patient will be sidebent left to engage the restrictive barrier. Please note that the same principles apply to treating flexion dysfunctions of the thoracic spine, except that the segment that is in dysfunction is where the thenar eminence is placed, specifically directly underneath the posteriorly positioned transverse process. The thrust is applied straight downward onto the segment (and your thenar eminence). The goal is to get the segment, in both cases, to move through its restrictive barrier so it is no longer restricted, and, thus, no longer in dysfunction.

23. C   HVLA is a direct technique. Thus, the patient must be set-up into his restrictive barrier. If the dysfunction is C4FRRSR, then the freedom is C4FRRSR; the opposite of the freedom is the restriction. Thus, the restriction is C4E/NRLSL. Assuming this is a typical case, we will work with C4ERLSL as the restriction. Thus, C4 must be placed into slight extension (anterior translation at C4 allows for extension to be undertaken), and the head and neck must be placed into a rotated left, sidebent left position in order to engage the restrictive barrier. Then, the thrust is a rotatory thrust in the leftward direction, towards the patient's eye, to move C4 through its restrictive barrier.

24. B   Ulcerative colitis is a disease of the colon. Diseased tissue launches the fight or flight mechanisms of the body, namely the sympathetic nervous system. The ascending and transverse colon are served sympathetically by T10-T11, while the descending colon and sigmoid colon are served by T12-L2. Such activation of sympathetics could cause inappropriate afferent impulses to be delivered to the T10-T11 spinal cord segments and the T12-L2 spinal cord segments, respectively. If these impulses are chronic in nature or particularly intense acutely, the threshold to activation of those spinal cord segments can be lowered, thus allowing for facilitation and associated outgoing impulses to other tissues and organs also innervated by those spinal cord segments. Thus, we have to be aware that a sympathetic response is very likely. Hyperovulation is not a sympathetic response; sympathetic responses are associated with physiologic responses that increase the likelihood for immediate, but not necessarily long-term, survival. Ovulation increases the likelihood for pregnancy, which is not consistent with survival interests of mother and baby. So, hyperovulation is a parasympathetic, and not a sympathetic, response. Decreased glomerular filtration rate (GFR) reflects decreased filtration by the kidney. Filtration of blood and, more importantly, production of urine (which also needs time to be excreted through urination, and also leaves a scent) is not consistent with IMMEDIATE survival. Thus, decreased GFR would be a sympathetic response. Increased vitamin B12 absorption implies improved small bowel, particularly terminal ileal, function. Absorption of nutrients is a rest and digest process (normal body function), a process that promotes long-term survival. Thus, sympathetics slow that process. Increased B12 absorption is consistent with parasympathetic stimulation. Menorrhagia is increased menstrual bleeding; it is excessive manifestation of a normal body function, and certainly, through the loss of blood and the promotion of a scent, not consistent with attainment of immediate survival. Thus, menorrhagia is a parasympathetic process rather than a sympathetic one. A decreased potential for urinary tract infections implies that there is no urinary retention, and less likelihood for contamination of the urethral opening. Sympathetic activation of the bladder causes urinary retention (cannot be taking the time to urinate if fending off a wild animal – and do not want to leave the scent). So, sympathetic activation, by way of causing urinary retention, actually increases the likelihood of UTI, since urinary retention allows for urine to stagnate and increases the likelihood for bacterial overgrowth before it is evacuated. Likewise, ulcerative colitis directly causes diarrhea, which increases the statistical likelihood for fecal contamination for the vulvar and urethral areas; so,

that can also increase the likelihood for urinary tract infections. Thus, decreased GFR is the only one of the options listed that would, physiologically, be a sympathetic response. The kidney is sympathetically controlled via T10-T11; so disease of the ascending colon and transverse colon would have the potential to cause decrease renal function.

25. A   An inferior anterior positioning of the sphenoid, palpated by the index finger, implies that the sphenoid is in flexion. This finding in this case is bilateral. Thus, the sphenoid is in flexion, which means that there is craniosacral flexion. This is supported by the occipital findings: the posterior aspect of the occiput is inferior and lateral. If there is craniosacral flexion, there is also external rotation of the paired bones of the skull. This creates a wide head, orbits that are widely displaced, and even a wide upper jaw, as the maxillae also externally rotated. The eyes appear prominent (bulging) bilaterally because the sphenoid dips forward, making them protrude outward. The ears appear prominent bilaterally because the temporal bone, a paired bone, rotates externally, allowing it along with the overlying ear on it to "stick out." Because of the temporal bone involvement with flexion (and extension) dysfunctions of the craniosacral axis, dysfunction of cranial nerve VIII is possible, yielding tinnitus, vertigo, or hearing loss. Beetle brow is the colloquial term for the appearance of the brow when someone has a near-vertical forehead; vertical foreheads are consistent with craniosacral extension, not flexion.

26. D   The dorsum of the foot is supplied by the L5 nerve root, and, more specifically, by one of its branches known as the common peroneal nerve. The patient demonstrates no features of disc or spinal pathology; hence, the pathology causing the dorsal foot pain is probably more distal, such as that involving the common peroneal nerve. A posterior fibular head can compress on the common peroneal nerve to produce sensory disturbance on the dorsum of the foot. Posterior fibular head will also, mechanically, force the foot to take on a position more consistent with supination: inversion ("pigeon-toed"), plantar flexion ("pointed foot"), and internal rotation. Thus, the patient's complaints are very consistent with a posterior fibular head.

27. D   During palpation, the right ASIS, right PSIS, and the right pubic bone are appreciated as being superior, meaning they feel as if they are more superior than the left. These are three landmarks used to determine innominate positioning. The right SI joint is tender; while that could mean pathology at the location of the right SI joint, it could also mean that there is tension being placed on that SI joint. Tension can be created by have the contralateral side of the pelvis either higher or lower than the one you are palpating. So, is the right innominate superior, or is it the left innominate that is inferior – giving the one the false impression that the right is higher. The standing flexion test is positive on the left, confirming that the true pathology is on the left. So, this is a left inferior innominate shear.

28. A   There is a deep right superior sulcus and a left shallow ILA; while there is a slightly shallow left superior sulcus, when viewing the mechanics of the sacrum, we can see that there is a definite tilt forward at the top right and a definite tilt upward at the bottom left. Down on one side at the top and up on the other side at the bottom is a forward torsion, the sacrum "twisting" forward around an axis. The axis is what it spins around, like a door on its hinges or like if there were a rod stuck into the sacrum and it were spinning around that. The axis represents the most normally positioned part of the sacrum; it may have anomalies, too, but those anomalies are not part of the torsion or spinning. So, these findings are consistent with a left axis. So, through palpation, it appears that there is a forward sacral torsion about a left axis. The seated flexion test is positive on the right. Flexion tests confirm the side of the pathology. Remember, the axis represents the most "normal" part of the sacrum. So, if there is a left axis, that means the bulk of pathology is on the right. The seated flexion test confirms this, and so also is a finding consistent with a torsion about a left axis. The patient complains of low back pain that is exacerbated by backward bending. Neighboring body regions move in opposite directions in order to maintain balance. So, if someone backward bends (extends their spine above the sacrum), the sacrum naturally moves into flexion. If it moved into extension with the rest of the spine, the center of gravity would be behind the person and would cause them to fall backwards. Thus, anytime someone backward bends, that forces the sacrum into flexion. If the right side is already dipped forward as part of the torsion, it will be forced into a more severe forward position with backward bending, promoting even more pain. So, this also is in support of a forward sacral torsion. Forward sacral torsions are always either right rotation on a right axis (R on R) or a left rotation on a left axis (L on L). Since we know that the axis is left, we therefore can surmise that the rotation is also left (because this is a forward sacral torsion). L5 can induce torsions; if it is responsible for the torsion, there are particular findings to be expected. L5 will be rotated in the opposite direction, and will be sidebent ipsilateral to the sacral oblique axis. Thus, since the sacrum is rotated left, L5 will be rotated right. Since the sacrum has a left sacral oblique axis, L5 will be sidebent left. Applying Fryette mechanics, if the sidebending and rotational components are opposite to one another, the segment must be in neutral. Thus, this could be associated with L5NSLRR.

29. A   The sphinx test is also known as the backward bending test, and is a test whereby the patient lies prone and propped up by their elbows, allowing the spine to be extended. The position resembles the sphinx, and so the name is applied. Remember that any position one assumes causes neighboring body regions to move in opposite directions in order to maintain balance. As we extend the spine above the sacrum, the sacrum moves into flexion. Thus, the sphinx test will naturally cause the sacrum to flex bilaterally, resulting in deep sacral sulci bilaterally. The purpose of this test, however, is to better differentiate between forward and backward sacral torsions, especially when the other palpatory findings are questionable. Without the sphinx position, we know that a forward sacral torsion results in one of the sacral sulci, either the right or the left, being deeper than the contralateral one. Likewise, without the sphinx position, we know that a backward sacral torsion results in one of the sacral sulci, either the right or the left, being more shallow than the contralateral one. Thus, in either case, there is asymmetry between the right and left superior sacral sulci; that is an expected finding for any sacral torsion. However, when the patient assumes the sphinx position, the sacrum will dip forward into flexion. That means that, if there is a forward sacral torsion, the originally deep superior sulcus will get even deeper, as it moves even further forward. However, there is a limit to which it can move like this and still be structurally part of the sacrum. The contralateral side, which represents the healthier side of the sacrum, will also move forward into a flexed stated. While the two sides will not assume the same level of "deepness" during the sphinx test, they will become less asymmetric (more symmetric). In other words, both will be thrust forward into flexion, the dysfunctional side still being deeper than the non-dysfunctional side – but the difference between them will be less distinct that when they were evaluated in a neutral position. On the other hand, if there is a backward sacral torsion, that means that one of the superior sacral sulci is dipping backward, and is RESTRICTED from moving forward (recall that dysfunctions are named for their freedom, and the opposite of the freedom is restriction) . So, when that patient is asked to assume a sphinx position, the healthier side with its superior sulcus will move into flexion while the dysfunctional side will be restricted from making any forward/flexive movement at all. Thus, the superior sulci with a backward sacral torsion will become more asymmetric (less symmetric) than what they were with neutral positioning of the body.

30. B   Positions of prolonged flexion can cause many problems, most distinct among them is an iliopsoas contracture (flexion contracture of the iliopsoas); this contracture can result in a full psoas syndrome, with many other low back and pelvic structures thrust into somatic dysfunction as a result to the contracture. The psoas is attached to T12-L5, so contracture of the muscle can affect any of those vertebral segments; L1 and L2 are the ones most frequently affected. When the muscle contracts, it pulls down on the affected vertebrae, causing them to move into flexion; they are also sidebent ipsilaterally due to the strain by the muscle. Accordingly, these vertebrae also assume an ipsilateral rotation. The psoas joins the iliacus within the true pelvis; the iliacus attaches to the iliac crest, iliac fossa, anterior sacroiliac ligaments, and the sacral ala. The iliopsoas shortening, from the contracture, causes the entire pelvis to shift contralaterally (sort of like sticking one's hip out to the opposite side). So, if there is a pelvic shift to the right, that could mean that there is a spasm bringing the lumbar spine close to the left pelvis, such as what would happen with a left iliopsoas contracture (left psoas syndrome). A left sacral oblique axis is also induced by these mechanical changes. The contracture also causes the body to "maintain balance" on the contralateral side by inducing yet another spasm: one of the piriformis. So, a compensatory result of a left iliopsoas contracture could be a right piriformis spasm. This can cause gluteal pain. It can also cause compression and/or irritation to the sciatic nerve, the right one in this case. That would exacerbate the right gluteal pain, and also cause right posterior thigh pain. Thus, it appears that the patient has a left psoas syndrome. As such, we can expect to find the lumbar vertebrae that are affected to be flexed (a Type II Fryette dysfunction; a non-neutral dysfunction), and rotated left, sidebent left.

REFERENCES

1. American Osteopathic Association, Foundations for Osteopathic Medicine, Williams and Wilkins, Baltimore, Maryland, 1997.

2. Beers, Mark H., The Merck Manual, Merck (Pub.), Rahway, New Jersey, 2006.

3. Cooper, Daniel H., MD, et al, The Washington Manual of Medical Therapeutics, Lippincott Williams & Wilkins, Philadelphia, PA, 2007.

4. Daffner, Richard H., MD, FACR, Clinical Radiology - The Essentials, Williams & Wilkins, Baltimore, Maryland, 1999.

5. DiGiovana, E.; Schiowitz, S., An Osteopathic Approach to Diagnosis and Treatment, J.B. Lippincott & Co., Philadelphia, Pennsylvania, 1991.

6. Dolinski, Lori A., MSc, PhD, DO. Boards Boot Camp Hi-Yield Compendium (7th Ed.), Pro-Medica Publishing Company, Ottsville, PA, 2009.

7. Fauci, Anthony S., MD et al (Eds), Harrison's Principles of Internal Medicine, McGraw-Hill, New York, New York, 2008.

8. Frohlich, Edward D., MD, MACP, FACC (Ed), Rypins' Basic Sciences Review, Lippincott-Raven Publishers, Philadelphia, PA, 1989

9. Gibbons, Peter, MB, BS, DO, DM-SMed; Tehan, Philip, DO, DipPhysio, Manipulation of the Spine, Thorax and Pelvis - An Osteopathic Perspective, Churchill Livingstone, New York, 2000.

10. Goldberg, Stephen, MD, Clinical Neuroanatomy, MedMaster, Inc., Miami, Florida, 1997.

11. Goldberg, Stephen, MD, The Four-Minute Neurologic Exam, MedMaster, Inc., Miami, Florida, 1999.

12. Guyton, A., Textbook of Medical Physiology, W.B. Saunders, Philadelphia, Pennsylvania, 1991.

13. Healey, Patrice M., MD; Jacobson, Edwin J., MD, Common Medical Diagnoses: An Algorithmic Approach, W.B. Saunders Company, Philadelphia, PA, 2000.

14. Hoppenfeld, Stanley, MD, Physical Examination of the Spine and Extremities, Appleton & Lange, Norwalk, Connecticut, 1976.

15. Howard, W.H. III, DO, Easy OMT, Momentum Press. Siloam Springs, Arkansas, 1998.

16. Klippel, J.H., Primer of Rheumatic Diseases, Arthritis Foundation, Atlanta, Georgia, 1997.

17. Kuchera, William A., DO. FAAO; Kuchera, Michael L., DO, FAAO, Osteopathic Considerations in Systemic Dysfunction, Original Works, Greyden Press, Columbus, Ohio, 1994.

18. Kuchera, William A., DO, FAAO; Kuchera, Michael L., DO, FAAO, Osteopathic Principles in Practice, Original Works, Greyden Press, Columbus, Ohio, 1994

19. Kumar, Vinay, MBBS, MD, FRCPath; Abbas, Abul K., MBBS,; Fausto, Nelson, MD; Aster, Jon, MD; Robbins & Cotran Pathologic Basis of Disease, W.B. Saunders Company, Philadelphia, PA, 2009.

20. Moore, Keith L., BA, MSc, PhD, FIAC, FRSM, Clinically Oriented Anatomy, Williams & Wilkins, Baltimore, Maryland, 1992.

21. Orient, Jane M., MD, Sapira's Art & Science of Bedside Diagnosis, Lippincott Williams & Wilkins, Philadelphia, Pennsylvania, 2000.

22. Pansky, Ben, PhD, MD, Review of Gross Anatomy, McGraw-Hill, New York, New York, 1996.

23. Rubin, Emanuel MD et al. (Ed.) Rubin's Pathology: Clinicopathologic Foundations of Medicine, 4th edition, Lippincott, Williams, & Wilkins, Philadelphia, PA, 2005.

24. Savarese, Robert G., DO (Ed), OMT Review, Robert G. Savarese, DO, New Jersey, 2003.

25. Sieg, K.W., PhD; Adams. Sandra P., PhD, Illustrated Essentials of Musculoskeletal Anatomy, Megabooks, Gainesville, FL, 1996.

26. Simmons, Steven L., DO, Osteopathic Manipulative Medicine, Steven L. Simmons, DO, Fort Worth. Texas, 2001.

27. Snell, Richard S., MD, PhD, Clinical Anatomy, Little, Brown, and Company, New York, NY, 1996.

28. Sobrin, Jack J., DO; Zaslau, Stanley, MD, Osteopathic and Musculoskeletal Medicine, FMSG Publishing Co, DeBary, Florida, 1998.

29. Taylor, Robert B. (Ed), Family Medicine Principles and Practice, Springer-Verlag, New York, New York, 1994.

30. Webster's New Complete Medical Dictionary, Merriam-Webster, Smithmark, New York, New York, 1995

31. Weissleder, Ralph MD, PhD, et al. Primer of Diagnostic Imaging, 3rd edition, Mosby, Philadelphia, PA, 2003.

32. Woodward, Thomas, MD, and Best, Thomas M., MD, PhD, The Painful Shoulder: Part I, Clinical Evaluation, American Family Physician, American Academy of Family Physicians, 2000

　　　　　　　　　Handbook of OMT Review

INDEX

Dalyrimple pump 130
Deep tendon reflexes 77, 94, 110, 156
Deformity 30, 86-90, 93
Delivery 51, 59, 61, 79, 147
Deltoid muscle 73-74, 76, 81, 91, 94
Dens 15, 19, 22, 30
DeQuervain's Tenosynovitis 85, 90, 159
Dermatome 20, 23, 42, 76, 94, 101, 110, 119, 121, 136-138
Descending colon 53, 115-116, 125, 128, 132-135
Diaphragms 28, 33, 52, 65, 69, 124-125, 128-131
Diarrhea 118, 120, 127, 136, 138
Direct technique 6, 8-9, 12
Disc
      Herniated 7, 19-20, 22, 35, 38, 43, 47-48, 91-92, 94, 110, 158-159
      Intervertebral 16, 19-20, 22-23, 35, 41, 43, 47-48, 91-92, 94
Diverticulitis 118, 164, 174
Dizziness 147
Dorsiflexion 100, 102, 105, 107, 110, 130
DTRs 77, 94, 110, 156
Eccentric contraction 6, 161
Edema 1, 103
Elbow 75, 77-78, 81-83, 93, 153-154
Epicondyle 81, 83, 97
Epidural nerve block 51
Erb-Duchenne palsy 79
Erector spinae muscles 36
Erythema 1
Esophagus 115-116, 128, 132, 134
Ethmoid bone 142, 146, 149
Extension 4-5, 11-14, 17-19, 21-23, 27, 38, 54, 56, 61-62, 66, 70-71, 77, 81, 83, 85, 87-90, 93, 97-98, 100, 125, 141-143, 146-149, 154
Extensor carpi radialis
      Brevis 77
      Longus 77
Extensor carpi ulnaris 77
Extensor digitorum
      Brevis 99
      Longus 103

Extensor hallucis longus 100, 103
Extensor pollicis brevis 85, 159
Extremity
      Lower 12, 41, 53, 56, 101, 130, 135, 153
      Upper 74-76, 83, 135, 157
Facilitated positional release 8
Facilitation 119, 121, 123-126, 134, 137-138, 156
Fascia 1, 8, 10, 12, 81, 83, 89, 124-127, 129-133, 156
Femoral epicondyle 97
Femoral head 96, 101-102, 109-111
Femoral head angulation 101-102, 109-110
Femoral nerve 53, 99
Femoral patellar grinding test 98, 155
Femoral patellar tracking syndrome 104, 108
Femoroacetabular joint 95-96, 111
Femur 52-53, 95, 97-98, 101-102, 110, 128
Ferguson's angle 36, 43, 47
Fibula 95, 97, 99, 102, 108-110
Finkelstein Test 85, 159
Flexion 4-5, 11-14, 17-18, 20-23, 27, 38, 41, 44, 48, 53-54, 56-71, 77, 81, 83, 86-90, 93, 97, 100, 102, 105, 107, 110, 130, 141-143, 146-149, 151, 153-155, 159
Flexion Tests
      Seated 56-64, 67, 71, 153
      Standing 56-64, 67, 70-71, 153
Flexor carpi radialis muscle 77
Flexor carpi ulnaris muscle 77, 81
Flexor digitorum longus muscle 103
Flexor hallucis longus muscle 103
Flexor pollicis brevis muscle 86
Foot 98-103, 105-108, 110, 130, 155, 159
Foramen
      Intervertebral 35
      Magnum 15, 141, 144
      Neural 41
      Sacral 51
      Sciatic 52
      Transversarium 15, 20, 23
Forearm 78, 81-83, 92, 125, 154
Fracture 7, 38-40, 49, 90, 93, 106, 148

Tibiotalar joint 105
Tinel's Test 85-86, 154
Tinnitus 147, 149, 151, 166, 177
Torsion 8, 31, 61, 63-65, 69, 70-71, 131,
       141-142, 144, 148-149, 151
Torticollis 161, 168
Translation 17-18, 21, 24
Transverse
       Arch 106-107
       Axis 5, 11, 13, 54, 141, 145
       Carpal ligament 85
       Colon 115-116, 125, 132, 135
       Ligament 15, 19, 22
       Plane 5, 13
       Process 4, 10, 25-26, 30, 32, 36, 66,
          70, 126
Trapezius muscle 16, 79
Trapezoid bone 84
Travell's trigger points 127, 156, 170
Triceps muscle 77, 81
Triceps reflex 77, 91
Triquetrum bone 84
Trochanter, femoral 53, 128
Tropism 42
Ulnar artery 75, 85
Ulnar nerve 74, 81, 86-90, 93
Umbilicus 42, 136
Uncinate joints 15
Uncovertebral joints of Luschka 15
Upper Extremity 20, 73-83, 85, 92-94,
      115,121, 125, 134-135, 153-154,
        157-158
Upslip 58, 66
Ureter 116, 132, 134-135
Uterus 69, 116-117, 134-135, 174
V Spread 147
Vagus nerve 113, 116, 118, 120, 133, 136,
      139
Valgus 82-83, 90, 93, 98, 102, 104, 108-110
Varus 82-83, 98, 102, 108-110
Vastus
       Intermedius 95
       Lateralis 95, 104
       Medialis 95, 104
Vault hold 147, 156
Venous sinus release 147

Vertical strain 141, 145
Viscero-somatic reflex 113, 123-124, 126-
      127
Viscero-visceral reflex 114, 118-119, 124,
      136-137
Whiplash 7
Winging of the scapula 79, 91, 93
Wrist 75, 77-78, 81, 83-86, 90-91, 93, 154,
      159
Yergason's test 78, 154
Zink, J. Gordon DO 131
Zygopophyseal joint 15, 25, 40
Zygopophyseal tropism 42

Handbook of OMT Review